CLIMATE
CHANGE
CYCLES

AND GLOBAL WARMING MYTHS

Charles S. Opalek, PE

This book is dedicated to my wife Mary, daughter Amy, and to all those martyrs of science who refuse to graze at the public trough of grant money, and who dedicate themselves to finding the real answers as to why our climate is changing through natural causes.

Acknowledgements:

Thank you to all those scientists, researchers, and writers listed in the bibliography for the plethora of information which made this book possible.   In particular, I acknowledge the work of two individuals; Theodore Landscheidt for his explorations into the effects of the Sun on the Earth brought about by the planets, and Edward R. Dewey who spent his entire life examining cycles of all types.

Disclaimer:

I am a sole practitioner; a self-employed consulting engineer. My clients are architects, contractors, builders and owners from whom all my livelihood is generated and appreciated. I have no financial interest in - nor do I want any support from – 'Big Oil', electric utilities, any other corporations, public officials, or private benefactors. My only goal is to replace dogma with science, emotion with measurement, and hysteria with induction.

Please direct all inquiries to the publisher: www.lulu.com

Contents

"The whole aim of practical politics is to keep the populace alarmed - and hence clamorous to be led to safety - by menacing it with an endless series of hobgoblins, all of them imaginary"

- H. L. Mencken

Foreword

My first book, <u>A Convenient Fabrication</u>, was published in 2007. It was the result of accumulating a vast collection of reference material gathered in responding to articles in trade journals and newspapers regarding global warming. I was getting fed up with all the hoopla about man destroying the planet with the burning of fossil fuels. Al Gore believed man was totally at fault for the warming. One popular radio talk-show host believed this phenomenon was all natural. As an engineer I knew the truth must lie somewhere in between. But how to quantify it? Then the realization struck me that I had collected enough researched material along the way to write a book.

My second book, <u>Wind Power Fraud</u>, was published in 2010. It looked into the subject of harnessing energy from the wind, only because the global warming crowd believed alternative energies would replace fossil fuels and eliminate manmade CO_2 from the atmosphere. But how efficient is wind power? One method is to analyze the amount of energy produced divided by the amount of energy that went into bringing these alternative energies to market. This method is known as the Energy Returned On Energy Invested, or EROEI. Finding EROEIs on alternative energy sources is not easy. I suspect the reason is: that if the truth got out about how unsustainable they are, these alternative energy projects would end up in the dumpster where they belong. After an exhaustive Internet search, the only EROEI analysis I found was for the Livermore Pass wind project in California. This project claimed to have an eye-popping EROEI of 15.83. Anything over a value of 1.00 is considered positive, meaning more energy will be produced than went into making it. However, after a line-by-line analysis of Livermore, I was able to show the claim of an EROEI of 15.83 was deeply in error and was more like a down-to-Earth value of 0.29, thus unsustainable.

This, my third book, <u>Climate Change Cycles</u> was needed for three reasons. One, to explore in detail where weather and climate have been and where they come from. Two, to analyze the cycles of weather and climate and determine their drivers with the aim of trying to predict with a relatively high degree of confidence into what direction the climate was truly headed. Three, to review the myriad claims that mankind is somehow responsible for 'climate change', when indeed anyone with a shred of common sense knows that the climate is always changing. The primary questions remaining about climate change can be boiled-down to these: by how much, when, and in what direction is the climate heading? The answers are in this book - and will astound you. Buckle up. You are in for an unexpected and exhilarating ride.

Charles S. Opalek, PE

> *"The aim of public
> education is not
> enlightenment at all, it
> is simply to reduce as
> many individuals as
> possible to the same safe
> level, to breed and train
> a standardized citizenry,
> to put down dissent and
> originality."*
>
> — *H. L. Mencken*

Introduction

Brief description of the book. The book starts off with the basics of climate mechanisms, environmental outcomes, the sun, the planets, and Earthly influences. Then it looks at the history of temperature, CO_2 and other records. Then various cycles are explored including those of the atmosphere, oceans, abundance, temperature, time, Sun, Moon, Earth, and planets. It then embarks on reconciling these various cycles in an effort to determine in what direction the future climate might be heading. Finally, it looks at the myths of global warming and climate change (as if it is only man that causes this), and the attendant scams.

Purpose of this book. There are over a thousand books available on global warming and climate change. Most of these books basically blame mankind for the alleged recent rise in temperature, due to burning of fossil fuels. The remainder attempt to show that it is natural causes that are responsible for the changes being witnessed. A small number attempt to predict future climate using very short periods of recent data to make projections far into the future. What I believed was necessary was a study of cycles produced by the influences of the celestial bodies around us, to see if there was another explanation of how our climate is controlled.

Journey of writing the book. Writing the first book was easy. Al Gore with his book, An Inconvenient Truth (AIT), gave me the template in which to operate. All I had to do was read his book and address his comments, observations and statements as they unfolded. Writing the second book was also somewhat easy. It only had to demonstrate that the actual efficiencies of alternative energies were a lot lower than advertised. It became easy when one particular EROEI analysis for a wind project, the proverbial needle in a haystack, was found. If it were not for this discovery, the second book probably would have never been thought of or written.

Problems with writing this book. As writing progressed, and chapters where identified and developed, the original outline of the book went into a state of disarray. Analyzing tidal, gravitational, and electromagnet influences of celestial bodies on the Earth's climate as drivers of climate appeared a daunting task and were getting nowhere. But when attention was directed at analyzing down-to-Earth measured, historical data, a clearer picture of what this book would become came into focus.

Research process. The process of gathering historical data, such as, temperature and CO_2, was easy. The process of finding scientific papers which came up with the equations detailing the effects of the Moon, planets, and Sun on temperature and CO_2 concentration proved somewhat more difficult.

Acknowledgments. If it were not for those scientists who developed the equations used in this book, which became the basis of my analysis of climate change cycles, this book would have never advanced to the point of where long range prognostications were possible.

How long. It took over eleven years to write Climate Change Cycles, as opposed to the first two, which collectively took much less time. This much longer period was the result of the work load of my consulting business and the difficulty with generating original material rather than tearing apart the published claims of others. It is a lot easier to criticize the work or others than generate original thought.

Chapter 1:
Climate Mechanisms

Climate is generally defined by many of the classical dictionaries as weather conditions prevailing at a particular place over long periods of time. The weather conditions enumerated include temperature, wind velocity and precipitation. The extent of 'a place' is not defined. The period of time is not specified. So what is climate and how does the climate change?

If the Earth's surface were flat, covered only by land, totally devoid of oceans, seas and lakes, had no ice caps, ice sheets or glaciers, rotated so slowly that the same spot on Earth always pointed to the Sun, and the Earth rotated around the Sun in a perfect circle, I suspect the weather and ultimately the climate on this planet would be static and very boring. This is the condition of planet Mercury; very hot facing the Sun, very cold on the opposite side as it rotates once on its axis and about the Sun at the same rate. Our Moon suffers from the same condition. This condition of the moon's surface always facing us, has caused this orbiting body to become distorted.

With no atmosphere, no hydrosphere, no cryosphere, there is essentially no weather or climate change on Mercury. However, if you move Mercury away from the Sun dropping its average temperature, increase its rotation from once a year to once a day, change its orbit around the Sun from a circle to an ellipse, add a moon that rotates around it every 28 days, add bodies of water that get tugged up and down by the Sun and its new moon, and you get climate. You get conditions like those on Earth with clouds, rain, snow, ice, evaporation, condensation, atmospheric oscillations, ocean circulations, tornadoes, hurricanes, storms, monsoons, droughts, etc. The fact that the Earth rotates on its axis as fast as it does creates the opportunity for impacting or even generating many of these changes.

Coriolis Effect

On the Earth circulation patterns are created in the atmosphere by the rotation of the Earth. This rotation imparts a Coriolis effect on the atmosphere, creating a differential torque or twisting action. This twisting action, coupled with the day-night heating of the atmosphere as it is alternatively exposed to the hot Sun, develops the longitudinal easterlies, westerlies, and polar air currents. It also produces the latitudinal Hadley cells, Ferrel cells and polar cells, which mix by overturning the warmer and cooler longitudinal currents.

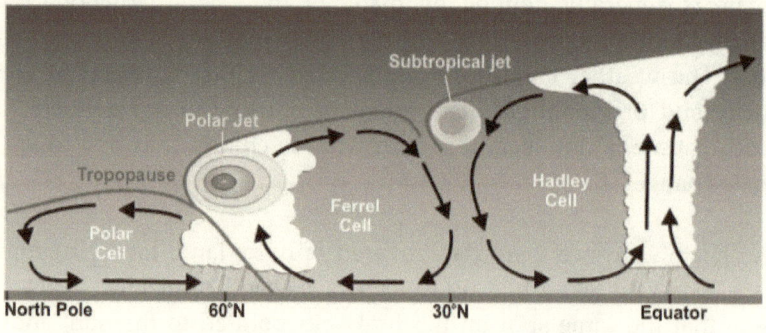

Fig 1a: Earth Diagram of Wind Currents
https://www.weather.gov/images/jetstream/global/jetstream3.jpg

The Coriolis effect is created by the rotation of the Earth and is quantified by the formula $2\ \omega \sin \varphi\ v$, where ω = rate of spin of the Earth, $\sin \varphi$ = latitude, and v = wind speed. Thus, the greater the wind speed or closer to the equator, the greater the Coriolis force.

The Coriolis effect strongly affects the large-scale oceanic and atmospheric circulations, leading to the creation of atmospheric features like jet streams and western boundary currents. Such features are in geostrophic balance; the Coriolis and pressure gradient forces balance each other. Coriolis acceleration is also responsible for the propagation of many types of waves in the ocean and atmosphere, including Rossby waves and Kelvin waves. It is also instrumental in the so-called Ekman dynamics in the ocean, and in the establishment of the large-scale ocean flow pattern called the Sverdrup Balance. But without an atmosphere or oceans, none of these effects would exist.

Vortices

Vortices manifest when wind created by Coriolis forces are deflected. These winds are deflected to the left in the Southern Hemisphere and to the right in the Northern Hemisphere. The stronger the pressure gradient the stronger the wind and the more deflection there will be. Vortices take the form of cyclones, tornadoes, hurricanes, typhoons, water spouts, and dust devils.

Most vortices are vertical in nature and the result of Coriolis forces. However, vortices often are created simply because of the interaction of fast moving upper horizontal winds being sheared by the stationary ground. These horizontal vortices become accelerated when winds roll down from mountains into valleys, sometimes creating mountainadoes.

Polar Vortex

Going back further to 1974, it was Time Magazine, that was proclaiming the beginning of the next Ice Age, blaming the polar vortex. Forty years later, it is Time Magazine again, but this time blaming the polar vortex for global warming. Not surprisingly, Time Magazine and the legacy media in general are attempting to get a lot of mileage out of the recent outbreaks of polar vortices and are placing the blame on man and global warming. But unlike shark attack stories which flood the legacy news for two weeks during the summer than disappear virtually overnight, outbreaks of polar vortices look like they will be around for a while. Time Magazine and their readers have had a convenient loss of memory, thus allowing two opposite scare stories from the same phenomenon within 50 years.

The polar vortex has its roots in the Arctic Oscillation, an oscillation associated with the Arctic Ocean <http://www.climate.gov/news-features/event-tracker/how-the-polar-vortex...>. But more on ocean oscillations later.

Polar Vortices

For decades before 2013, there was little mention of the "polar vortex". Now, we not only hear about the polar vortex in the winter, but even in other seasons. Suddenly, there is an escalation in the appearance of this lens of very cold air dropping out of the Arctic. Or is there? Let us take a deep breath and review the facts.

The figure below shows the extremes of polar vortices. The reference to "noaa" is the National Oceanic and Atmospheric Association, which is a regulatory agency of the US Department of Commerce..

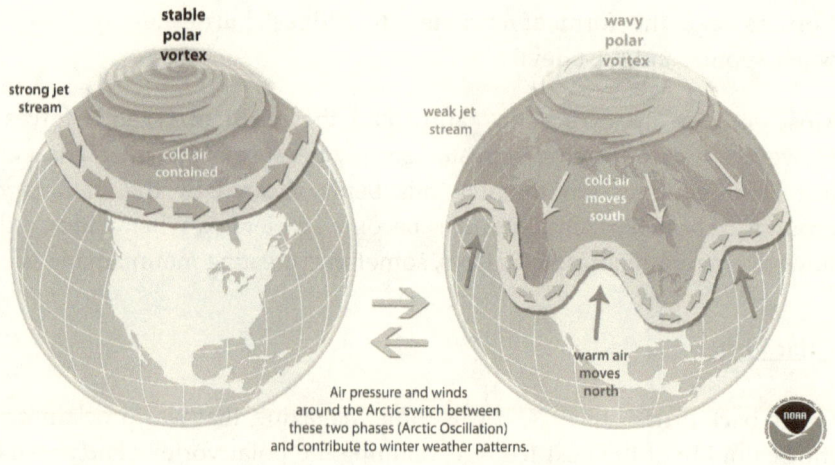

Fig 1b: North Polar Vortex
https://www.noaa.gov/image_download/3913?itok=5yZosn1E

Polar vortices are nothing new. They manifest not only in the Arctic, but Antarctic as well. Actually, a distinction should be drawn here. Two manifestations are at work. One, is the circumpolar vortex in the Arctic and Antarctic. Two, is the development of mobile polar highs. Let's look at these differences.

Circumpolar Vortices

Circumpolar vortices (CVs) are high speed winds that travel west to east at about 20 degrees from the poles. The Arctic CV rotates mostly over land, around the Arctic Ocean. The Antarctic CV rotates over the Southern Ocean around Continental land.

During their respective winters these vortices grow very strong. The Antarctic CV becomes so strong, it virtually prevents the ingress of ozone rich atmosphere. This blocking action, together with the plummeting production of ozone from lack of Sun results in the famous ozone hole, peaking in size and decreasing density in January. By July, with a return of the Sun and diminishing strength of the Vortex with attendant reentry of ozone-rich atmospheres from the tropics, the ozone hole decreases in size and increases in density.

The Antarctic CV is very stable, because it hovers over water. The Arctic CV is unstable, because it is greatly affected by changing geography of the land mass it travels over. This instability is a primary factor which creates the ability of cold pockets of air to escape and travel southward out of the Arctic. These escape paths are not random, as you will discover later on.

Mobile Polar Highs (MPH)

MPHs are outbursts from the polar regions towards the equator. Mobile polar highs just don't occur randomly anywhere. They are primarily controlled by topography. In the northern hemisphere they usually head south over the Hudson Bay lowland, Scandinavia, and Bering Sea. In the Southern hemisphere they usually head north toward the west coast of Chile, northeast past the tip of the Cape of Good Hope, and northeast across Australia. Mobile polar highs regularly surge out of Canada into the Great Plains of the USA.

There are three main trajectories or MPHs in the Arctic: the Scandinavian, the Pacific and the American. It is the American which is receiving most of the attention. Between 1989 and 1993, there were a total of 328 MPHs recorded. That's an average of about 2.27 MPHs per month per trajectory. This data comes from the Laboratory of Physical Geography <http://www.applet-magic.com/MPH.htm>.

Palmen-Newton model of global circulation

Palmen-Newton is a modeling of general circulation of the atmosphere from the equator to the poles. This is a perfect example of heat transfer. Energy is transferred from the hot equator to the cold poles. It starts at the equatorial, intertropical convergence zone, works its way thru the trade winds, subtropical jet, roaring forties, polar jet, and polar easterlies.

13

Along this path are the four usual cyclonic air patterns of moonsonal lows, subtropical highs, extratropical lows and polar highs.

Changes in the Jet Stream paths

The Jet Stream is four regimes of fast moving air about 1-km deep, 100-km wide. There is the North Polar, North Temperate or Subtropical, South Temperate or Subtropical, and south Polar Jet Streams. The two northern jet streams travel easterly and two southern jet streams travel westerly. Along their paths they veer North and South forming three to four Rossby waves.

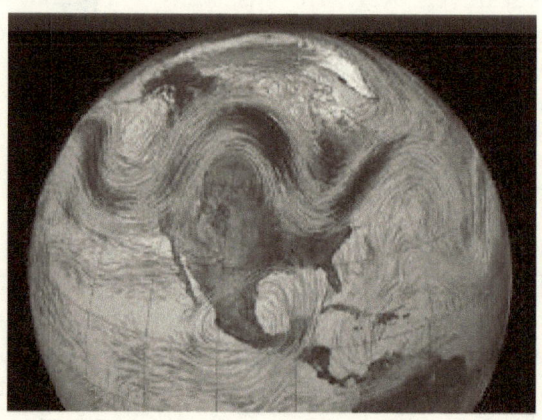

Fig 1c: North Polar Jet Stream
https://svs.gsfc.nasa.gov/vis/a010000/a010900/a010902/COVER_1024x576.jpg

These wave patterns have summer and winter characteristics. During winter, the Jet Stream is positioned over the southern USA allowing the cold arctic air to enter. During the summer, the Jet stream usually is positioned over southern Canada allowing tropical air to enter the country. When the Jet Stream wanders between these two positions, especially during the spring and fall, more transient conditions occur which cause more frequent changes in weather. These waves propagate from west to east above the equator and east to west below the equator. These waves, when not that pronounced, allow the movement of high and low pressure areas from west to east, as usual. However, when Rosby waves get really distorted, sometimes even backtracking on themselves, heat waves or cold waves lasting for days can be the result. This is particularly true of high pressure cells, which are then referred to as blocking highs.

*"Science never solves
a problem without
creating ten more."*

– George Bernard Shaw

Chapter 2: Environmental Influences

As the Sun's weather changes it not only changes the Earth's weather it actually alters the chemistry of the Earth's atmosphere.

The atmosphere experiences mass transport as expressed in circulations and oscillations. The atmosphere experiences heat transfer where cooler areas receive heat from warmer areas. The atmosphere experiences diffusivity where volumes of air in one area mix with and alter volumes in other areas. The atmosphere is also not a homogenous gas with constant properties regardless of location. Of significance, water vapor enters and leaves various areas of the atmosphere. The atmosphere is also affected by other outside stimuli. Most of these come from the Sun. Some come from outside the solar system. Many of these rays are stopped by the atmosphere, but some make it thru to the surface of the Earth. These influences include cosmic rays, gamma rays x-rays, and ultraviolet radiation. Each of these rays can alter the chemistry of the atmosphere.

Electromagnetic Spectrum

Each of the aforementioned rays occupy a specific range of frequencies. To get a picture of their range and location with respect to each other, take a look at the electromagnetic spectrum below.

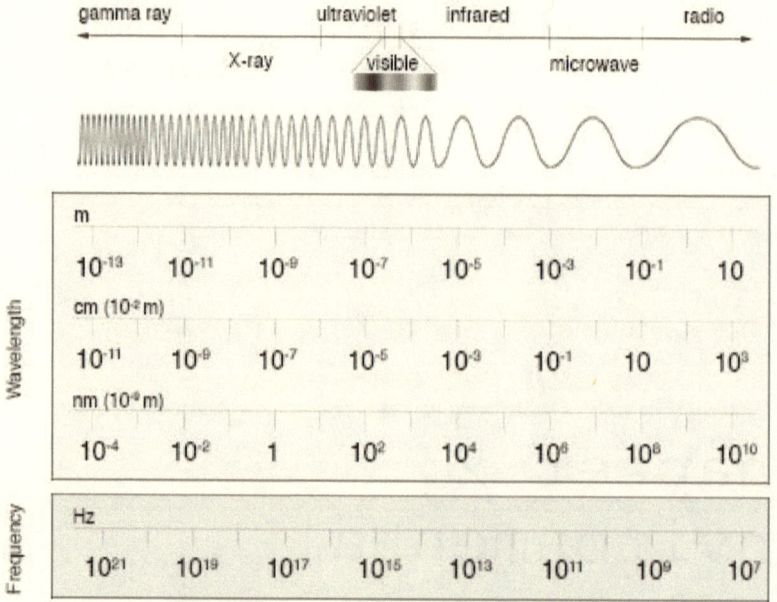

Fig 2a: Electromagnetic Spectrum
<imagine.gsfc.nasa.gov/science/toolbox>

The amount of energy contained in cosmic rays is higher than all other types of electromagnetic propagation. Cosmic rays though not shown above are to the left of gamma rays and have a wavelength of E-21m, or about 100,000,000 times more potential energy than infrared radiation at E-14m.

The amount of energy, h, is proportional to the frequency, f, and is governed by the equation:

$$h = c \times f, \text{ where c is Planck's Constant 6.63E-27 erg·sec.}$$

The lowest form of energy is conduction, followed by radiation and finally direct ionization. At the very bottom of the scale is infra-red radiation. The long waves from the sun are converted to short waves after bouncing off the Earth. This residual radiation is then partially responsible for creating the wind, which is barely capable of being converted into electricity by wind turbines. The ultimate conversion factor from radiative energy to mass transport of air is embarrassing low.

Cosmic Rays

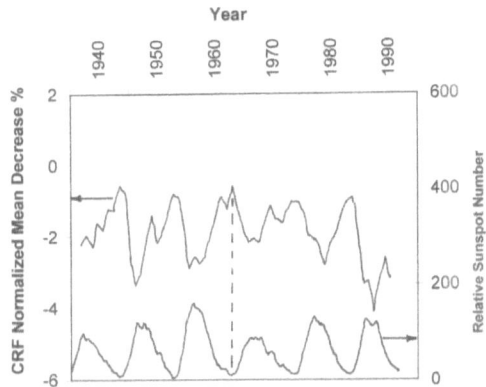

Fig 2b: CRF & SSN for Last 60 years
<http://www.drsi.dk/~hsv/Noter/solsys99.html>

Above is a plot of Cosmic Ray Flux and Sunspot Number between 1935 and 1995. The dashed vertical line shows one of the correlations between these two parameters during the period. The odds of this not being a random coincidence is 2^11, or 2047 out of 2048, or 99.95%. It is obvious: The more sun spots there are with their attendant solar wind, the lower the cosmic ray flux.

Gamma Rays

After cosmic rays, gamma rays are the most active form of radiation. Fortunately for us, gamma rays are rendered ineffective as they pass through the nitrogen-rich atmosphere. Nitrogen molecules when they are bombarded by gamma rays are split into nitrogen atoms and the gamma ray is extinguished. Thus, for each otherwise chemically inert nitrogen molecule that is split, two nitrogen atoms become available to enter into chemical reactions.

X-Rays

Next in the progression of energy capacity are X-rays. They do not effect nitrogen molecules, but do bombard oxygen molecules. The oxygen molecule is split to form two oxygen atoms, which in turn are made available to enter into chemical reactions.

X-rays generated by the sun come in bursts. They are classified as big X-class, medium M-class and C-small class. X-class flares are known for causing world-wide radio blackouts and long lasting radiation storms. M-class flares can cause brief radio blackouts near the Earth's polar regions. C-class flares cause few noticeable consequences on Earth.

UV Radiation

Finally, there is ultraviolet radiation (UV) which when colliding with oxygen molecules will split them into oxygen atoms. These atoms are then free to combine with oxygen molecules and create ozone. Ozone acts as a shield to UV radiation.

In the first few months of 2004 extreme weather conditions and a high solar wind combined to kill off 60% of the atmosphere's ozone molecules <http://www.nature.com/news/2005/050228/full/news050228-12.html>.
The reason for this reduction was traced to a 4-fold increase in nitrogen oxides, which are known for being ozone killers. I don't recall any burst of CFCs being released that could have caused this dramatic reduction in ozone. As to where this nitrogen oxide burst came from, I suspect was solar in origin.

Unlike gamma and x-rays, UV rays do reach the surface of the Earth, but in amounts generally safe for life. So, keep on that sun screen to protect your skin and UV filtering sunglasses to protect your eyes.

Spectral Absorption Bandwidth of Earth's Atmospheric Gases

One of the ways that the performance of high quality audio amplifiers was determined was to measure the output signal amplitude to a constant input signal over the operating frequency range of the unit. The most faithfully reproduced music comes from amplifiers with a constant output over the range. This is bandwidth. In a similar fashion the effectiveness of certain atmospheric gases to absorb energy from the Sun can be measured. Some gases absorb more energy than others.

Below is a pictorial showing the relative absorption of energy by nitrous oxide, methane, oxygen and ozone, carbon dioxide, water vapor, and the total of all these gases. This pictorial is distorted on purpose by the wavelength of absorption being plotted logarithmically. If this data were plotted linearly, much of the detail that occurs in the first two-thirds would get compressed and almost become unreadable. Plotting the data logarithmically alleviates this problem. For example, the right-third of the area under the curve for water vapor from about 10 to 100 Angstroms (1 Angstrom = 1 millionth of a meter) is about 90 percent of the total from 0.1 to 100, but only looks like about one third of the area. For oxygen & ozone, for example, the area from 0.1 to about 0.5 represents only about 4% of the entire bandwidth, but looks much larger, about one-fifth of the bandwidth.

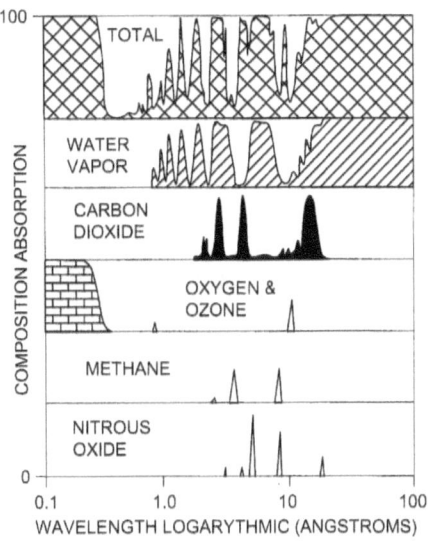

Fig 2c: Spectral Absorption Bandwidth for Atmospheric Gases
<http://www.barrettbellamyclimate.com/page15.htm>

There is much hoopla over the energy absorbed by carbon dioxide, CO2. But hardly anything is said about the energy absorbed by water vapor. Referring to the graph above, the bandwidth absorption of energy by water vapor is impressive, covering the range from 1 to 100 Angstroms. Looking at the sliver of absorption by CO2 looks puny in comparison. If you compare the area under their respective curves, the energy absorption by water vapor is about 26 times greater than that of CO2. The energy absorbed by the other gases is insignificant.

Water Vapor

All the discussions about global warming revolve around the concept that greenhouse gases absorb short wave radiation. There is never any mention of how much, if any, incoming long wave radiation is absorbed.

Incoming long wave solar radiation is absorbed by water vapor. Many references make the claim that no radiation is absorbed by incoming long wave radiation.

The amount of absorption has been determined to be 13.0 W/m2 (Tarasova and Fomin, Journal of Applied Meteorology, Vol 39, Issue 11, Nov 2000). Again, the estimated radiative effect of CO_2 created since the Industrial Revolution is 1.5 w/m2. Assuming the latter to be true, the effect of water vapor's incoming long wave radiation absorption is 13/1.5 or 8.66 times as great as CO_2's short wave radiation absorption.

Water Vapor vs CO2

The total weight of water vapor in the atmosphere is 1.558×10^{14} tons. The total weight of CO_2 in the atmosphere is 2.996×10^{12} tons. Thus, there is 52 times more water vapor in the atmosphere than there is CO_2. If you multiply the H_2O/CO_2 weight ratio by the of H_2O/CO_2 energy bandwidth ratio of 26 you get 1,352. Therefore, the amount of solar energy absorbed in the atmosphere by water vapor is about 1,352 times greater than that absorbed by CO_2. Said another way, if CO_2 increased from 315 ppm in 1958 to 409 ppm in 2017, that 30% increase in CO_2 could be wiped out by a 0.577% increase in water vapor.

Greenhouse Gas Effect (GGE

Only 3.27% of all CO_2 generated comes from man, the other 96.73% comes from nature. Only 0.001% of water vapor comes from man; the other 99.999% comes from nature. Water vapor by a factor of 26 has more of a spectral absorption bandwidth or GGE than does CO_2. After adding the contributions of methane, nitrous oxide, and CFCs it turns out only 0.28% of the GGE comes from man, the other 99.72% comes from nature. If man ceased to exist, the reduction in the GGE would be hardly noticeable.

But why be bogged-down with evidence of centuries, millennia, and epochs showing that the recent minor variations in temperature and climate pale in comparison to the past? It is for this very reason the global warming, climate change crowd look only slightly into the past and project minor changes today into the future. If the climate and temperature changes of hundreds of millions of years were to be splattered on the headline news, the whole climate change, global warming industry would fall flat on its face.

Ozone

You would think that all the hubbub about ozone and the infamous Ozone Hole was something that came about recently. But actual ozone measurements of the atmosphere officially began in 1927 in Arosa, Switzerland.

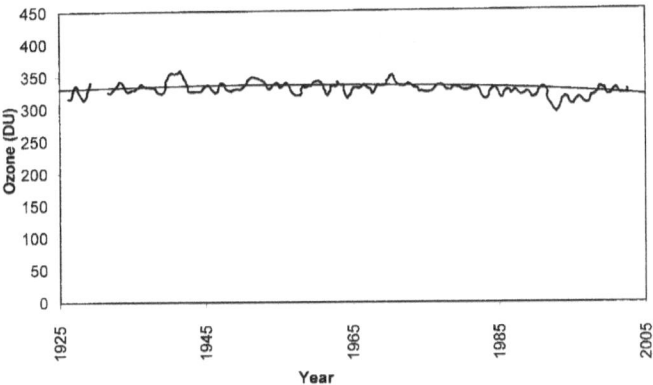

Fig 2d: Ozone Concentration, Arosa, Switzerland 1926 to 2001
(based on http://www.johnstonsarchive.net)

From the graph above, simple observation tells us that not much has happened to the ozone concentration over the last century, wobbling around between 300 and 350 Dobson Units.

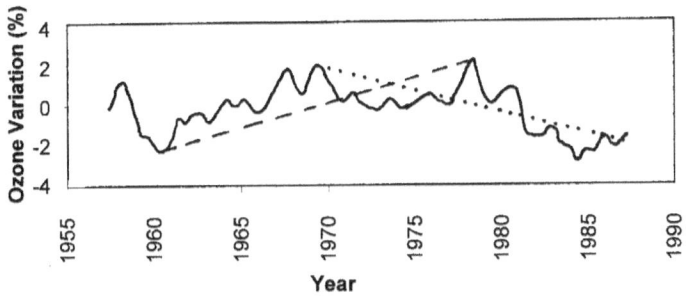

Fig 2e: Global Ozone Variation, 1958 and 1987.
(adapted from Angell, J. K., On the Relation between Atmospheric Ozone and Sunspot Number, Journal of Climate, 01 Nov 1989)

However, that didn't stop the National Aeronautics and Space Administration (NASA)'s Ozone Trend Panel (OTP) from cherry-picking the Fig 2d innocuous, graph. The OTP initiated the Ozone Hole Crisis proclaiming an ozone depletion trend of about 4% within the time frame between 1969 and 1986 as depicted by the dotted line in Fig 2e above.

Had the OTP used the time frame between 1960 and 1978 as shown by the Fig 2e dashed line above, an ozone _**increase**_ of 4% would have been the result. However, cherry-picking data is important in order to generate a crisis. So, the OTP time frame of 1969 and 1986 is the official ozone depletion trend and the entire basis used by ozone hole alarmists.

<u>Roland and Molina</u>

In 1995 Frank Sherwood Rowland and Mario Molina were awarded the Nobel Prize in Science. Their theory was that CFCs (Chloro Fluoro Carbons) produced by man when reaching the upper atmosphere, would be broken down by the Sun, releasing chlorine, and thus create a hole in the ozone layer. This theory and its numerous chemical reactions, with as yet undetermined reaction constants, have never been demonstrated in the laboratory, or for that matter found to actually exist in the upper atmosphere.

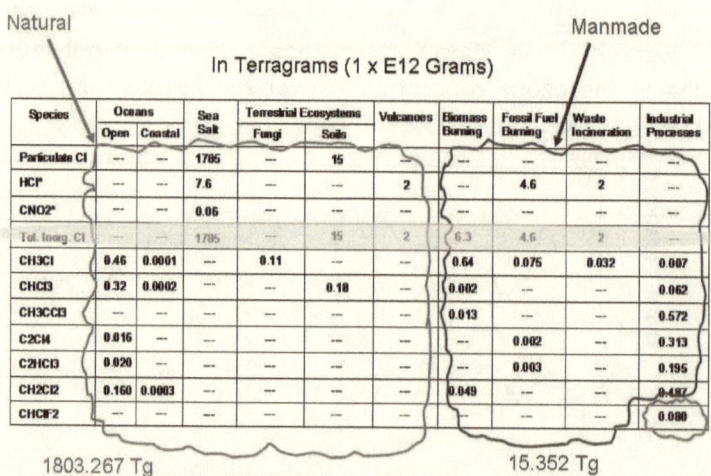

Natural   Manmade

In Terragrams (1 x E12 Grams)

Species	Oceans		Sea Salt	Terrestrial Ecosystems		Volcanoes	Biomass Burning	Fossil Fuel Burning	Waste Incineration	Industrial Processes
	Open	Coastal		Fungi	Soils					
Particulate Cl	---	---	1795	---	15	---	---	---	---	---
HCl	---	---	7.6	---	---	2	---	4.6	2	---
CNO2	---	---	0.06	---	---	---	---	---	---	---
Tot. Inorg. Cl	---	---	1795	---	15	2	6.3	4.6	2	---
CH3Cl	0.46	0.0001	---	0.11	---	---	0.64	0.075	0.032	0.007
CHCl3	0.32	0.0002	---	---	0.18	---	0.002	---	---	0.062
CH3CCl3	---	---	---	---	---	---	0.013	---	---	0.572
C2Cl4	0.016	---	---	---	---	---	---	0.002	---	0.313
C2HCl3	0.020	---	---	---	---	---	---	0.003	---	0.195
CH2Cl2	0.160	0.0003	---	---	---	---	0.049	---	---	0.407
CHClF2	---	---	---	---	---	---	---	---	---	0.090

1803.267 Tg   15.352 Tg

Fig 2f: Sources of Chlorine
<www.eurochlor.org/rcei>

I always thought it was unfair to blame chlorine from CFCs for causing the Ozone Hole. While on a cruise ship at sea, I could not help being overwhelmed by that characteristic aroma of the ocean - the smell of chlorine. Seeing the wind chop off the crest of waves and sending ocean spray into the air, it prompted to me to look into how much chlorine is put into the air by oceans versus how much might come from man. To get the answer, I figured why not go to the chlorine industry and see if they have any data on chlorine sources. After a little research I stumbled upon the table above on sources of chlorine on our planet.

From the above table it can be seen that 1803.267 Terragrams (Tg) of chlorine come from natural sources, with 1705 Tg coming from the sea salt of Earth's 329,000,000 cubic miles of oceans. A Tg is equal to 1,102,311 tons, or about 5 times the weight of the Oasis Of The Seas cruise ship. Of the 15.352 Tg of chlorine that is manmade, only 0.08 Tg come from CFCs ($CHCF2$). Dividing the total of 1803.267 plus 15.352 by 0.08 you get 22,372. That is, there are 22,372 times more chlorine atoms coming from natural sources and manmade sources other than from CFCs.

So, how does the ozone layer above the Antarctic have the intelligence to discriminate the chlorine atoms from man from the 22,372 atoms available from all sources, that allegedly destroy the ozone molecules that protect us from ultraviolet radiation? The answer is: It doesn't and it can't.

Roland and Molina were given the Nobel Prize not for any work in chemistry, or the Nobel's basic conditions for awarding the prize (ie: "for outstanding achievements in science that leads to industrial progress and the benefit of mankind"). They were awarded the Nobel Prize for the "political implications that saved mankind from an impending catastrophe". So there you have it. The Nobel Prize has nothing to do with science. It's all about politics, derived from an imaginary hobgoblin.

Summary

- In w/sf, clouds have 18 times more influence than CO2

- In w/sf, water vapor has 8.66 times more influence than CO2.

- The amount of solar energy absorbed in the atmosphere by water vapor compared to that of CO2 is 1,352 to 1.

Chapter 3: Environmental Outcomes

Our environment consists of four scientific spheres: Atmosphere (air), hydrosphere (water, water vapor and ice), lithosphere (land), and biosphere (living organisms). Within these four spheres various climate anomalies can materialize, including melting glaciers, rising sea levels, stronger storms, more severe droughts, more devastating floods, and bigger forest fires.

Melting Glaciers

Melting glaciers which presumably cause sea level rise are another favorite of the warmistas (global warming fanatics). They like to point to a handful of glaciers around the world which appear to be receding rapidly, thus proving their case of global warming. However, what they fail to mention is the glaciers around the world which are growing. To be objective the only true representation of world glacier conditions is to examine the entire world inventory of glaciers.

The warmistas also seem to be oblivious to the fact that glacier calving is a natural occurrence, happening with regularity. This is evidenced by deposits found at the bottom of the ocean, which come from the originating location of the glaciers. These cycles are referred to as Ice Rafting Debris cycles.

Continental glaciers and ice fields, Greenland continental glaciers and Antarctic continental glaciers make up about 0.63% of the total inventory of the world's cryosphere. The remaining 99.27% of the world's inventory of ice in Antarctica and Greenland is solidly grounded in a deep freeze and isn't going anywhere.

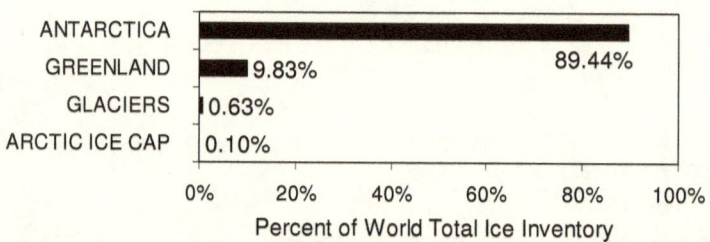

Fig 3a: World Total Ice Inventory
<www.johnston archive.net/environment/waterworld.html>

Yes, many glaciers are melting, but there are a significant number of glaciers which are also growing. Out of an estimated 170,000 glaciers, about 71,000 have been inventoried, of which only 246 have any mass balance data available. If mass balance data exists for only 246 glaciers out of the 71,000 inventoried, how can any claim be made that *almost all* of the mountain glaciers are melting?

In 1946 there were mass balance records on only three glaciers. From 1956 to about 1968 the number of mass balance records increased from 10 to about 82. Since 1968 the average number of mass balance records has remained about 78.

Of the few glaciers that have been surveyed, most are of the maritime variety, which are easily accessible. Not surprisingly, these are the glaciers which have the most melting since they are closest to the waters towards which they are heading. In many cases their noses are literally sticking into - and floating on - water. Of the continental glaciers, which are the most difficult to access, some are growing and not receding.

Of the total distribution of ice inventory throughout the world, the Arctic Pack Ice is a mere 0.10%. This pack ice averages about 2.5 meters thick and covers an area on average of about 9,000,000 square kilometers. This is analogous to a piece of paper three one thousandths of an inch thick covering an area the size of four football fields. The fact that this fragile, almost two-dimensional, layer of ice at the North Pole doesn't disappear and reappear on a regular basis is astounding. And yet the Artic Ice Pack hangs in there resisting the claims of its demise as it has since the 1960s.

The Greenland Ice Cap accounts for 9.80%. Recent news stories relate how the Greenland ice is melting. Yes, on the edges there appears to be some melting. However, what is taking place inland is a totally different story.

On July 15, 1942 during WWII, 'Glacier Girl' along with five other Lockheed P38 fighter planes and two B17 bombers made an emergency landing on an ice field on the Greenland interior. All the crewmen were rescued, but the planes were left there, abandoned. In 1992, after searching for 50 years, Glacier Girl was recovered. It was found buried under 268 feet of ice. That is not a misprint – we're talking 268 <u>feet</u> not 268 inches. So, the next time someone tries to tell you about the ice in Greenland melting, tell them to look up the 'Glacier Girl'. <www.p38assn.org/glaciergirl/index.htm>

The mother lode of ice inventory is the Antarctic Ice Cap/Sheets/Shelves which contain 89.81%. Recent news stories and scientific articles are coming out about how chunks of ice the size of Manhattan are breaking away from the ice shelf dangling out into the Barent Sea giving the impression of catastrophic melting. Yes, there is melting in this area, but it is due to Southern Ocean currents, underwater hydrovents and volcanic activity beneath these ice shelves. What is happening inland, again, is another story. Over the last five decades, the average temperature at the South Pole has been decreasing. Further, scientific stations at the South Pole get continually buried under new snow and ice and have to be rebuilt. The British, tired of having their facilities buried, have built a station on skis which can be towed around and kept on top of the ice and snow. So much, for Antarctic ice melting.

While on the subject of melting glaciers, I cannot pass up the opportunity to silence the global warming critics on one of their favorite 'canaries in the coal mine' - the Arctic Ice Cap. We are told since the start of satellite sea ice monitoring in 1979 that there has been a drastic decline in sea ice extent beginning in 1988. This is true. However, what these alarmists forget to tell you is that there was a similar loss of sea ice extent beginning in 1902 up to 1935. More significantly, between 1936 to 1987 there was a huge increase in the sea ice extent.

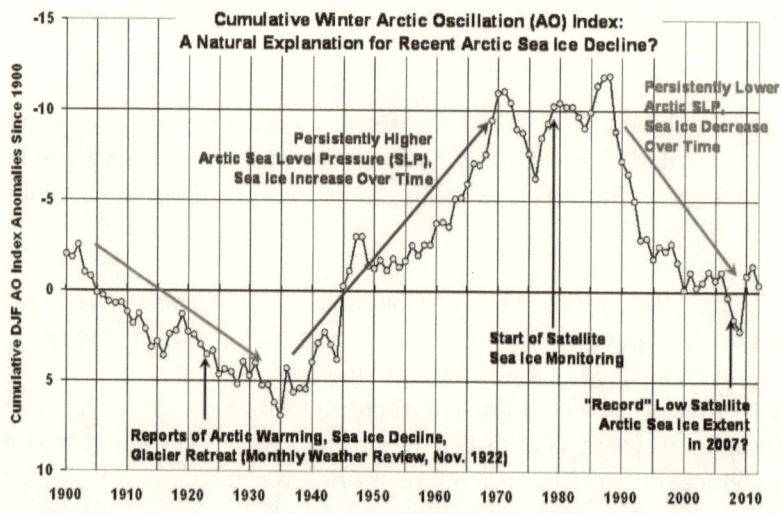

Fig: 3b Arctic Sea Ice Extent DJF – 1900 to 2013
<https://wattsupwiththat.com/2012/03/26/sea-ice-news-volume-3-number-3/>

Let's look above at the Arctic Sea Ice Extent during December, January and February (DJF) since 1900. Looks like a potential 74 year cycle to me. And notice the 1971 and 1988 double top. This will be discussed in a later chapter regarding climate forecasting.

The subject of melting glaciers, ice, etc., can be wrapped up in the resulting change in sea level. Traditionally, sea level change is quantified by Sea Level Equivalent or SLE. According to antarcticglaciers.org, an SLE of 1 mm change in sea level is equal to 360.87 Gigatons (Gt) or 393.7 km³ of ice. Assuming the world's total ice inventory of 29,960,000 km³ lost 9,150 km³ since 1960, the resultant increase in sea level would be 23 mm <www.johnstonarchive.net/environment/waterworld.html>. This equates to about 0.38 mm/yr, or 15 one thousandths of an inch, roughly five times the thickness of a human hair.

Rising Sea Levels

Sea level rise is a favorite of the warmistas. This gives them the opportunity to present their before-and-after pictures of what the coastline of Florida, New York City and similar low lying land masses would look like with sea level rise of a few dozen feet. What the warmistas hope you don't discover is that over ice age periods, sea level changes about 400 feet! Of course, this fact is conveniently forgotten when it seems recent changes of a few millimeters per year should be considered alarming, but changes of hundreds of feet in the past should be ignored.

Rising sea levels is another of the more common consequences of supposed manmade global warming. Let's pretend for a moment that this is true. It is estimated that 20% of the sea level rise measured during the twentieth century was from melting alpine glaciers. Assuming that sea level rise is increasing at the rate of 1.85 mm/year after 100 years sea level should have risen 185 mm. So, if the other 80% of alpine glaciers melt, the additional rise should be 740 mm. Thus, the total sea level rise from all melting alpine glaciers would be 925 mm, or about 3 feet.

With all the alpine glaciers gone, where would any further increase in sea level come from? If the Arctic Ice Cap melted there would be no sea level increase since this Cap is floating in water. Yes, there is always some talk about Greenland's ice melting and floating out to sea. But this is an impossibility, since almost all of Greenland's ice is trapped within its saucer-shaped geology in a permanent deep freeze. The Antarctic Ice Sheets, which by definition are on land, are for the vast inland part always below freezing, so no melt water is available here. The Antarctic Ice shelves are by definition out over the water, so any melt water from here would not result in any change of water level, either.

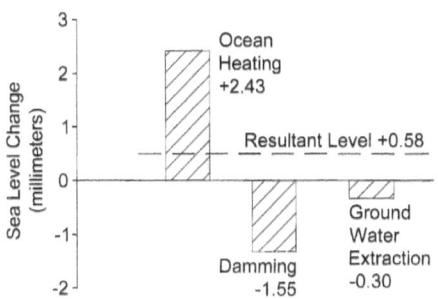

Fig 3c: Sea Level Rise

29

Besides melting alpine glaciers, there are many other, much more influential, factors influencing sea level change as shown above. There is thermal expansion from ocean heating. SeaLevelRise.Org estimates a 6.5 inch rise due to the oceans heating since 1950. That is 165.1 mm over 68 years or a 2.43 mm/yr increase. Man has an impact on sea level change, but it has nothing to do with burning of fossil fuels. Damming of rivers has been estimated to cause a _decrease_ in sea level of about 1.55 mm/year . Man's influence by groundwater extraction which causes land subsidence is estimated to cause a sea level _decrease_ of 0.3 mm/yr. (Bryant, Edward, Natural Hazards, Cambridge University Press, Second Edition, 2005).

According to SeaLevelRise.org about 2/3s of the sea level rise is due to ice melting and 1/3 is due to ocean thermal expansion. The slowing down of the Gulf Stream could raise sea levels at the USA east coast by 1 to 3 feet. So, you can see there is a lot more going on regarding sea level rise, than simply the melting of a few paltry glaciers.

One method by which sea level rise is determined is from tidal gages spread mostly around the coasts of the world. Not all tidal gages are measuring increases; many are measuring decreases. Globally, the average annual variability of sea level is 35 mm. Thus, the hypothetical sea level rise of 1.85 mm/year is being determined from a variability about 19 times greater than the supposed increase. All tidal gage data is extremely 'noisy', requiring mathematical manipulation and filtering. I have a problem with this. Who are the people doing the filtering? What agenda do they have?

Another more recent method of measuring sea level rise is by the TOPEX/POSEIDON satellite. The estimate of sea level rise based on satellite measurement is consistent with that of tidal gages, 1.85 mm/yr. But are the orbits of satellites constant without any change whatsoever? Are the land masses on which tidal gages and the satellite orbits from which altitude measurements are being taken stable? It has been estimated that the city of Moscow rises and falls 20 inches twice a day in response to the Moon's passing over. Where in this scheme of all this measuring is there a reference point that is "fixed" from which we can confidently say what is really going on? I don't think anyone has those answers.

In 100 years at 1.85 mm/yr rate of increase, it appears that sea level will rise 7.28 inches. I think New York City and Florida can rest easy.

<u>Storms</u>

Without question when it comes to a poster boy for climate change Superstorm Sandy would appear to be in front for the nomination. How could a storm of such immense size cause so much destruction, all from a hurricane that barely made Category 1 status? Sandy should have been a footnote in hurricane history; just another hurricane which took a northeastern trajectory and fizzled out in the cooler waters of the North Atlantic. But a series of events occurred which would forever change the lives of millions and cost over $100 billion dollars in damage.

First, Sandy started taking on extra-tropical characteristics. Instead of having a wind field a couple hundred miles across like most hurricanes, it developed a wind field of about 1,100 miles in diameter. Its cloud extent was much larger. Take a look at the satellite photo below from the Geostationary Operational Environmental Satellite (GOES). If Sandy was positioned a thousand miles further south, as it was two or three days earlier, it could have easily covered the entire eastern seaboard of the United States.

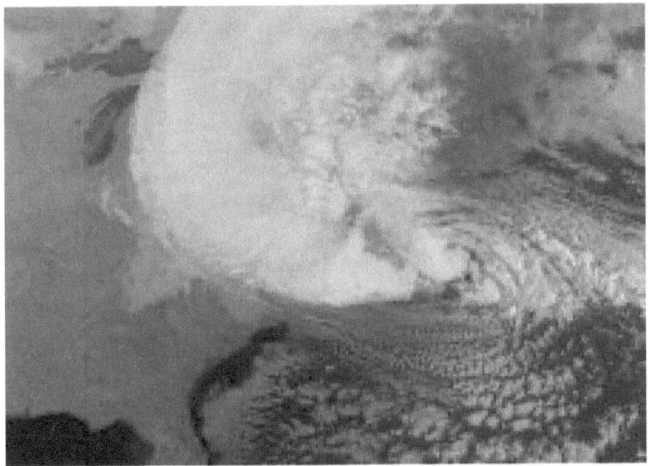

Fig 3d: Hurricane Sandy viewed from space
(NOAA/NASA GOES Project)

Second, there was a blocking high over Greenland, which would ultimately put the brakes on Sandy's normal trajectory. Instead of heading northeast, Sandy was stopped, turned sharply to the west, and taking the path of least resistance headed for the northeast coast of the USA. Third, to Sandy's west was an approaching upper level trough of cooler air. This temperature extreme provided additional energy, much like the dynamics of cold air rolling down from the Rockies into warm moist air coming up from the Gulf of Mexico, the recipe for making tornadoes in Tornado Alley in the nation's midsection. When Sandy took on extratropical characteristics in addition to its tropical origins, it became a hybrid storm. Fourth was the full moon. When Sandy made landfall, the high tide came in 6-feet higher than normal. Fifth, was the storm surge. At battery Park in lower Manhattan the surge was measured at 13-feet. As the Atlantic Ocean was pushed into lower New York Harbor, the surge of water simply had no place to go but up.

Sandy was also a warning long in the making. For decades, marshlands and other coastal low lands which acted as buffers to storms were being developed, placing millions of people in harm's way. It was only a matter of time before a catastrophe like Sandy would happen. Manhattan island hundreds of years ago was a buffer for the mainland. For centuries the map of lower Manhattan has increased in area with landfill from coal ash and trash. Lower Manhattan today is twice the area it was four hundred years ago. The West Side Highway, FDR Drive, and entrances for the Battery Tunnel are all located in this new manmade real estate. On October 29, 2013, after centuries of encroachment by man, Nature decided it was time to reclaim what is rightful its own.

Though Sandy got a lot of press, storms the size and destructive power of Sandy have occurred numerous times in the recent past, which apparently escaped the attention of the MSM (Main Stream Media – the legacy mediums of newspapers, magazines, and TV/radio broadcasting). There was the Halloween Storm of 1991 which was also a perfect storm, the Ash Wednesday Nor' Easter of March 1962, the Great Atlantic Hurricane of 1944, and New York Hurricane of 1893. Sand deposits analyzed, which were left by the Norfolk and Long Island Hurricane of 1821 reveals a much larger storm surge than Sandy, but since the hurricane hit at low tide there was little impact. Christine Brandon a graduate student at Amherst who analyzed this storm said, "even though Hurricane Sandy was the biggest event that happened in our lives, it is about par with things that happened in the past" <www.nbcnews .com/science/science-news/superstorm-sandy-exposes-new-jersey-m..>

To put things into perspective about how unusual Sandy or Katrina was, remember the bigger picture about hurricanes, typhoons, and cyclones. The annual Accumulated Cyclonic Energy or ACE, a measure of the total energy available within these storms has slightly increased over the last 50 years, but the frequency of tropical cyclones has slightly decreased. Yes, there are exceptions, and sometimes the exceptions are monsters.

Droughts

To listen to the global warming crowd, you would be lead to believe that droughts are something new and connected to man. In one sense droughts are nothing new, they have been around for millennia. In another sense, droughts can be connected to the activities of man, but not for the reasons you think.

Most of the great civilizations of the past were based on agriculture, and were dependent on consistent rainfall to ensure successful yearly harvests and a steady food supply for their societies to flourish. The downfall of most of the great civilizations of the past can be traced solely or partially to periods of drought ranging from decades to centuries in length. Great civilizations including the Harappan, Olmec, Aksumite, Minoan, Mycenaean, Mayan, Anghor Wat, Kymer, Nabta Playa have met abrupt ends because they could not cope with extended periods of drought.

Most of the developed countries have learned from the mistakes of the past. They have avoided over-cropping and land clearing, and embraced soil management practices. Today, developed societies are more prepared to deal with drought which comes with normal changing climate. As recently as the 1930's these practices were not in place in the American Midwest and resulted in the infamous Dust Bowl.

Droughts march to the tune of oceanic oscillations. The Indian Dipole Oscillation controls the drought and flood conditions surrounding it including Africa, India and its neighbors, Indonesia and eventually Australia. The Pacific Decadal Oscillation (PDO), El Nino Southern Oscillation (ENSO), and North Atlantic Oscillation (NAO) have also been identified as drivers of drought cycles. Droughts have been linked to the 18.6-year Lunar Cycle. Droughts in the Northern Great Plains recur in cycles of about 160 years. Around 2000, the PDO "flipped" resulting in the start of the record breaking drought in California. The historic drought in California persisted until the "Pineapple Express of 2016" abruptly ended it with drenching rainfall.

Then there are the pictures of Lake Mead and Hoover Dam showing the monumental decrease in water levels. This is not due to drought conditions. These plummeting water levels are driven, more and more, by people and industry tapping into the limited resources of the Colorado River which feeds these bodies of water. In fact, today the Colorado River barely reaches its historic destination of the Gulf of California due to all the water being siphoned off by humans.

And finally this: In an article in the Barrier Miner, January 23, 1933, <u>Yo-Yo Banned in Syria, Blamed For Drought By Moslems.</u>

<u>Floods</u>

Of the world's 27 largest floods, the 11 worst occurred more than 10,000 years ago (USGS Circular 1254) of the United States Geological Survey. That 10,000-years-ago figure is not a misprint. Long before the advent of fossil fuel burning, the world saw floods with estimated peak discharge rates up to 18 million cubic meters or 4.755 billion gallons per second, far beyond the puny floods being witnessed in modern times. But how should the severity of a flood be measured? Loss of life? Loss of property? Loss of commerce?

The following is a list of the world's 10 worst floods as measured by death toll over the last thousand years <www.epicdisasters.com> :

1	Huang He River, China	1931	1,000,000 to 3,700,000
2	Huang He River, China	1887	900,000 to 2,000,000
3	Huang He River, China	1938	500,000 to 900,000
4	Huang He River, China	1642	300,000
5	Banqiao Dam, China	1975	230,000
6	Yangtse River, China	1931	145,000
7	Netherlands & England	1099	100,000
8	The Netherlands	1287	50,000
9	The Neva River, Russia	1824	10,000
10	The Netherlands	1421	10,000

Fig 3e: World's 10 Worst Floods

The top 6 all time world's worst recent floods as measured by loss of life occurred in China. Please note 5 of the 10 worst floods happened before the Industrial Revolution, and the 2nd worst flood happened in 1887, just after the Industrial Revolution got started.

Then there is flood insurance. This taxpayer financed security has allowed people to take on risky behavior that logic would tell them to avoid. "So, what if my house is in an area prone to flooding, which has destroyed the lives of thousands of others. I have flood insurance. I can rebuild, again. And insurance will pay for it!" But why would you? Either you have little respect for yourself and your loved ones, or you are comfortable in knowing someone else is subsidizing your stupidity.

Increasingly, floods in populated areas are being aggravated by the decisions of man. This can take the form of development for housing by filling in areas of rivers that previously functioned as natural surge ponds for rising rivers. Levies have also been constructed to protect populated areas from river flooding. However, the addition of levies only shifts the problem of rising water upstream and downstream for others to contend with.

How many millions of dollars have been spent in New Jersey replenishing the sand dunes after a destructive Nor' Easter, so people can enjoy the luxury of living in their summer residences - living on-the-edge with no regard for potential catastrophe - all at someone else's expense?

Forest Fires

Forest fire activity remains at the mercy of lightning strikes, underbrush stockpiles and human interferences. Regardless of how dry a forest gets, there cannot be a forest fire unless there is a source of ignition. In Europe, only 2% of the known causes of brush fires is from natural causes such as lightning. Over half are the result of arson, and 40% from human carelessness. (Bryant, page 146).

Humans are the greatest cause of fires. Many such fires are started by children, most of whom are boys playing with matches. The size the resulting fires can be is usually determined by whether they are started by arson or lightning. In the case of arson, fires are conveniently started from nearby roads or public access campsites, from which fire fighting forces can quickly snuff them out. In the case of lightning, however, fires usually are in much more remote locations, which provides time for these fires to become conflagrations, not easily accessible by fire fighting forces.

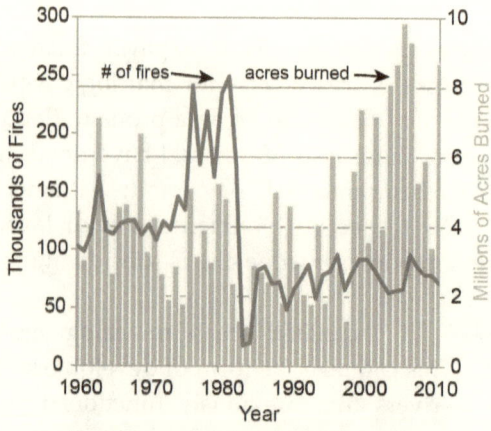

Fig 3f: Number of US Forest Fires
<nca2014.globalchange.gov>

The graph above is Figure 34 from the 2014 National Climate Assessment Report (NCAR) and depicts the number of forest fires from 1960 thru 2010. From 1960 to 1972 there were about 120,000 forest fires per year. From 1973 to 1982 the number of forest fires peaked to about 260,000 per. Two years later the number of forest fires plummeted to about 25,000 per year. Since 1985 the number of forest fires has stayed within a range of 50,000 to 75,000 forest fires per year. Contrary to the crisis mongers, there hasn't been any increase in the number of forest fires.

Wild Fires

Man has also interfered with nature. Natural fires tend to burn off the tremendous inventory of underbrush, twigs, leaves, and branches or limbs deposited by trees due to insect infestation, wind, or just old age, from which forest fires get their fuel. Many mega-fires are the result of "controlled burns" that got out of control. In many locations urbanization has encroached upon the forests. In Southern California it is common for new developments to become surrounded by naturally growing chaparral, a highly flammable plant, offering abundant high-energy fuel for fires. Man has for decades interfered with nature's way of maintaining balance, by suppressing fires, that would normally clear out much of this fuel. By suppressing fires, inventories of fuel increase to levels where, if ignited and gotten out of control, could develop into firestorms, which may not be easily tamed.

Logging practices have, also, contributed mightily to wildfires. Legislation passed in California has prevented the logging of forests. Thus, laws passed to placate environmentalist's concerns about plundering our forests for lumber had the unintended consequences of providing abundant fuel reserves for future wild fires.

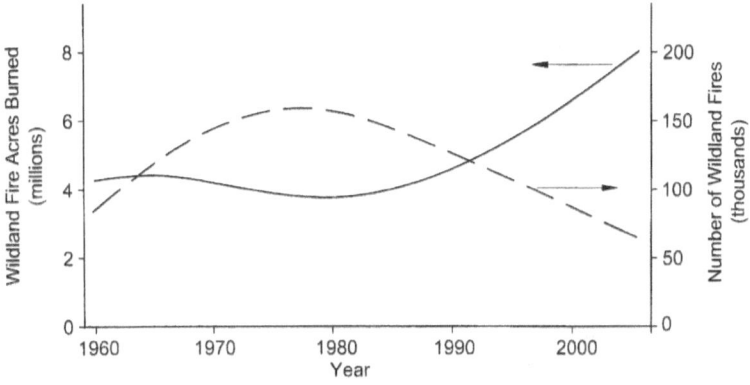

Fig 3g: US Wildland Fires: 1960-2006
<ncdc.noaa.gov/img/climate/research/2006/...>

The National Climatic Data Center (NCDC) graph above reveals that dividing the acres burned by the number of wildland fires is about 40 acres per fire. Also, while the acreage burned has increased since about 1988, the number of wildfires have decreased about 50% since 1978.

According to the USDA Forest Service, forest Heath Protection, Fig 16-1, the total acreage burned in the US around 1920 was 10 million acres, peaked at about 50 million acres in 1930, and decreased to about 5 million acres between 1960 and 2000. Since 2000 there has been a slight increase to about 9 million acres in 2008. The conclusion reached here is that for the last 60 years the acreage burned stayed very low compared to the previous 30 years.

Possible Catastrophes

Everyone knows that stars sometimes blow up. These are novas, and don't happen frequently. Hopefully, it is not in our Sun's near future. Yet, history shows our Sun's apparent consistency of operation has had some miniscule hiccups along the way.

Soil samples around the world have revealed micro particles of glass called micro-tektites. These are tiny glass beads measuring 0.1 to 1 mm in diameter. They have been found in Texas, Georgia, the Chesapeake Bay, Ivory Coast, Libya, Central Europe, Wabar Saudia Arabia, and dozens of locations around Australia, the Philippines, and nearby Indian Ocean. <www.dieholdfoundation.com/collecting-samples-of-sediments.html>. These particles have also been found in Lunar samples taken by Astronauts on the Moon. Could these particles have come from volcanic eruptions? Or more likely, could they have come from mini-novas produced by our Sun?

There are some who subscribe to the theory that ice ages begin violently and end suddenly. Historical data suggests these events happen in just a few decades and even as low as a few years. If this were true even an extraordinary above average accumulation of snow during successive winters could not explain the rapid buildup of ice over a mile thick which has happened in previous ice ages. To reach an ice thickness of 5,280 feet, it would have to snow about 52,280 feet (assuming a 10-to-1 snow to water conversion) or 633,600 inches. If history is correct with Ice Ages occurring over 5 (few) years, the snow would have to come down at the rate of 14.5 inches per hour. Thus, it is not hard to picture how the frozen remains of Wooly Mammoths were found with them standing upright and stomachs full, as if they were caught in a snow downpour.

Summary

- Based on the ice melting since 1960, the resultant increase in sea level would be 7.28 inches in the next 100 years.

- Stronger Storms are not borne out by the decrease in Accumulated Cyclonic Energy (ACE) over the last 40 years

- Droughts experienced recently are nothing compared to the civilization killers of the past, and march to the tune of oceanic oscillations.

- Of the world's 27 largest floods, the 11 worst occurred more than 10,000 years ago.

- Forest Fire activity remains at the mercy of lightning strikes, underbrush stockpiles and human interferences, not CO_2.

"A belief which leaves no place for doubt is not a belief; it is a superstition."

- Jose Bergamin

Chapter 4: The CO2 Record

Mauna Loa – CO2 from 1958 to the Present

Without question the boogeyman of global warming is carbon dioxide. Every K-8 student, housewife, factory worker, cab driver, everyone from every walk of life, knows this as if it were an immutable fact. Or, so you would think according to a supposed consensus of scientists feeding at the immense trough of free public research grant money. Al Gore, Hollywood actors, the legacy media, academia, the green movement, environmentalists, etc., all believe CO2 from the burning of fossil fuels is the singular cause of global warming.

Since 1958, the start of the International Geophysical Year, CO2 measurements have been conducted atop Mauna Loa, in Hawaii. In foresight, this location out in the middle of nowhere was thought to be an ideal location, unaffected by industry, transportation and population. In hindsight, this is probably the worst location that could have been chosen. Mauna Loa is atop the world's largest active CO2 belching volcano and surrounding by the world's largest CO2 belching ocean. The resulting curve of this measurement is below.

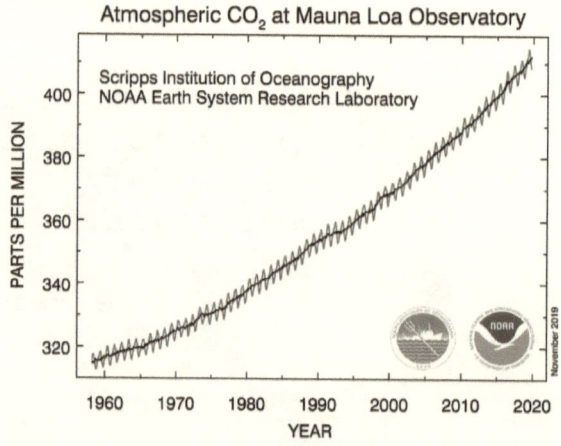

Fig 4a: Mauna Loa CO2, 1958-2020, 320-400 ppm
(Courtesy of NOAA)

The upward slope of this curve looks very ominous, as if things are changing very rapidly. This is due to the starting point of the curve anchored at 315 ppm, rather than the starting point being zero. This is a common technique to conserve space of presentation. This is also a common deceit to present data graphically as being more threatening than it really is.

Drawn with a vertical scale starting with 0-ppm it looks as follows:

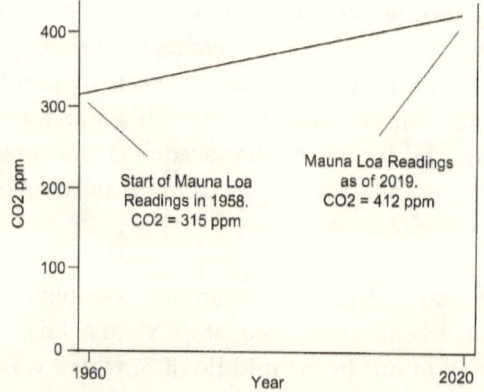

Fig 4b: Mauna Loa CO2, 1958 to 2020, 0-400 ppm
(based on NOAA ESRL curve above)

This is the same data but using zero as a drawing reference for the y-axis. Suddenly, the NOAA ESRL (Earth Sciences Research Laboratory) curve doesn't look that significant. But, you say, the increase from 315 to 412 comes out to be over 30%. This percentage is significant. Certainly this kind of change in such a short period of 61 years cannot be dismissed as something natural. It must be due to man. Right? Well, not really.

CO2 From 1820 to the Present

Here is another curve over a period from 1820 to 2000.

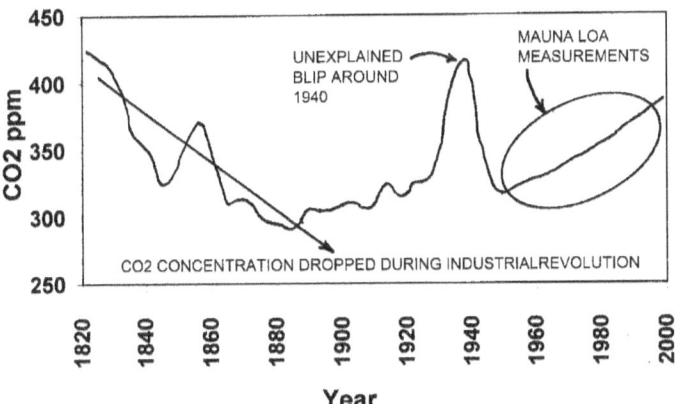

Fig 4c: CO2 Measurements 1820-2000
(Beck, Ernst-Georg, Energy & Environment, Vol 18, No. 2, 2007)

You can see the infamous Mauna Loa curve as the last 25% of this curve. The beginning period between 1820 and 1880 takes us thru the height of the Industrial Revolution. You would think with all the associated coal burning that occurred during this period that CO2 would be on the rise. However, just the reverse is true. While the air was being filled with choke-filled smoke from coal burning between 1820 and 1880, the CO2 concentration declined. Most notably, in 1820 and again in 1940, CO2 reached levels exceeding those of today.

CO2 For The Last 550,000,000 Years

But why look back only a paltry 56 years at the Mauna Loa data, or 180 years as documented by Beck. Let's look back 550,000,000 years.

The graph below is devastating to the global warming crowd. In geological terms the miniscule Mauna Loa CO2 data is barely discernable at the end of the above curve. With 550,000,000 years of data it is hard to see anything else except a most detailed and historically accurate CO2 record of the past. This is the all-time largest curve of data, not a ludicrous 100-year projection of where the paltry 56 years of Mauna Loa CO2 data is heading.

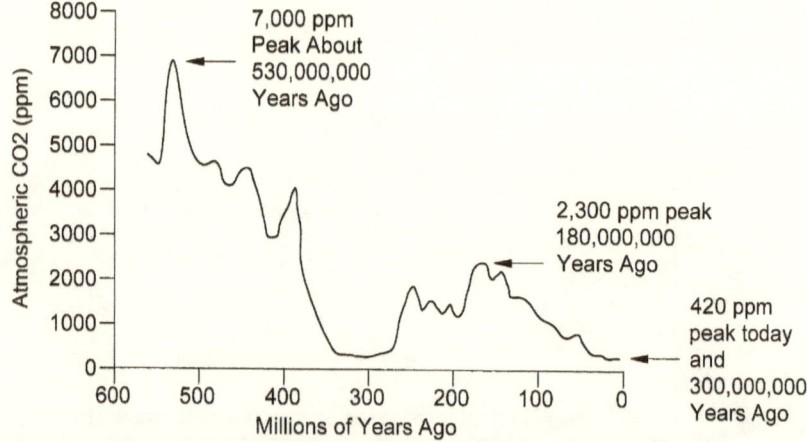

Fig 4d: CO2 Measurements for the last 550,000,000 Years
(Berner and Kothavala, AJS, **301**, February 2001, page 197)

A scientist who will remain nameless once said with respect to global warming from CO2 increases, that we are walking toward the edge of a cliff in the fog not knowing where the edge is. To this I say: Baloney. We know – and he knows, or should know – exactly where we are. We are 6,600 ppm below the all-time CO2 high of 7,000 ppm. We are at 400 ppm CO2, or 94.3% below the all time record high. We started out in 1958 at 315 ppm or about 4.5% above the all-time record low. Since that time CO2 concentrations from the perspective of hundreds of millions of years has moved from 4.5% to 5.7%, with respect to the all-time record low. We have advanced 1.2% toward the all time record high which is still 94.3% away. So, what's the problem?

CO2 for the last 1,700 Years

If you look back millions of years, any short term cycles there are would be obliterated. If you look back a few dozen years, the data would not be sufficient to develop short term cyclical trends. So, is there any data sets longer than 180 years, but shorter than hundreds of millions of years? The answer is, yes. Analyzing stomata cells found in fossils of plants as a climate proxy we can determine what the CO2 levels have been for over the last 1,700 years.

Critics may question the CO2 measurements using stomata cell data. I will cut the critics some slack here on the magnitude of the CO2 measurements. I will concede that there are significant statistical errors which could be introduced, which could lead to erroneous conclusions. However, what is of importance is the *cycles* of CO2 that are being recorded, not the measured concentrations.

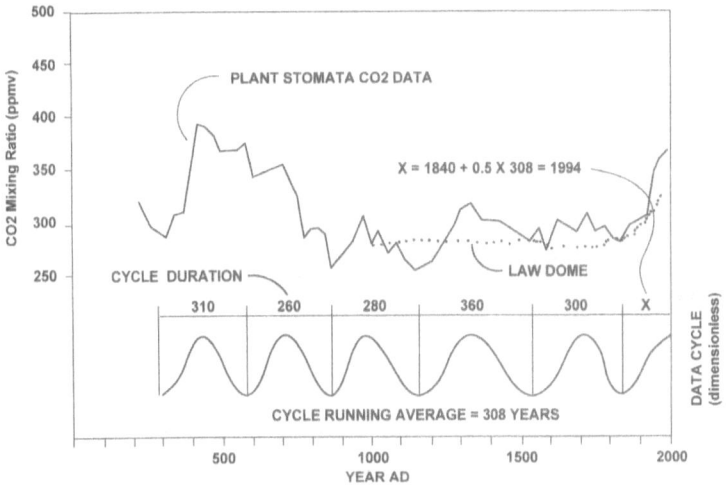

Fig 4e: 308-year cycle of CO2 measurements, 300-2000AD
(Kouwenberg et al)

First, look at the dotted line above of the Law Dome ice core data. This is the common graphical representation of where CO2 levels have been over the last millennium. There isn't much movement of CO2 levels over this period. The reason for this rather stable reading of CO2 concentrations is because ice cores do not easily record sudden changes of CO2 concentration over a period of decades or centuries.

Now, look at the plant stomata CO2 data above. Plant stomata CO2 measurements are much more detailed and show sudden changes. Plant stomata CO2 concentrations have cycled between 250 to 400 ppmv of CO2 throughout the last 1,700 years. This fluctuation occurred without the impact of an Industrial Revolution, major fossil fuel burning, or other influences from mankind. So, how would the crowd at the IPCC (International Panel on Climate Change) explain these fluctuations? They cannot. So, the global warming, climate change crowd sticks with the Law Dome ice core data, rather than explain shorter term cyclical changes they wished did not exist, can't explain, or simply wish to ignore. The significant, short term CO_2 changes registered by plant stomata are totally lost in the CO_2 changes registered by ice cores.

The significant conclusion that can be drawn from the above is earthshaking. Virtually everyone today believes CO_2 concentrations being reported have only been increasing. What the plant stomata demonstrate is that CO_2 concentrations are cyclical. What the cyclical curve above shows is that between 300AD and 1840AD there have been 5 distinct cycles of CO_2 concentration between 260 years and 360 years in duration, averaging about 308 years. If we take one half of this cycle of 154 years and add it to the low point of 1840, we arrive at the next peak of 1994. Taking into consideration the aforementioned durations, it becomes obvious that today's CO_2 cycle is at or near its historical peak.

Is it any wonder that a statistical pause in temperature began in 1996? If history is any teacher, CO_2 concentrations will begin their descent very shortly. Can you imagine the horror there would be in the world of Mann, Hansen, Jones, Briffa, Trenberth, Schmidt, et al, when this scenario begins to unfold? They already are dumbfounded to explain "the pause" in temperature. What will be their explanation for a CO_2 decrease? It's no mystery. We already know what the answer will be. They will turn the argument for manmade global warming being caused by fossil fuel burning into the argument for global cooling caused by fossil fuel burning. It is man's fault no matter what!

CO2 Changes since 2002 from Satellite

When we hear of CO_2 measurements it is assumed the numbers we hear are spread uniformly around the world. Well, guess again.

On May 2, 2002 NASA launched its Aqua Satellite. One of six instruments aboard was the Atmospheric Infrared Sounder. AIRS is part of the Earth Observing System which characterizes the entire atmosphere from top to bottom for surface and temperature, atmospheric temperature and humidity profiles, cloud amount and height, and the spectral outgoing infrared radiation.

As a result of these measurements it is possible to view the global temperature, carbon dioxide, carbon monoxide, sea level, ozone, and water/ice conditions of the Earth. These views are available NASA's website at <http://eyes.jpl.nasa.gov/eyes-on-the-earth.html> All you need to do is download the App onto your computer.

After running the program you will be able to select CO2 and get a color image of the CO2 concentration from 360 to 410 ppm (parts per million). The following image is the black and white version of what you will see.

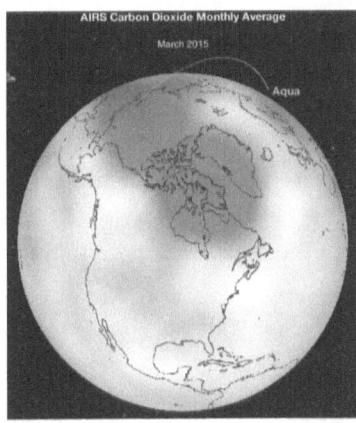

Fig 4f: Aqua Satellite AIRS CO2 March 2015 Average
<http://eyes.jpl.nasa.gov/eyes-on-the-earth.html>

In the application, you can point to anywhere on the Earth and get the ppm to within 2 decimal places. The view above has a range of about 397.7 (white areas) to 402.6 ppm (dark gray areas), or almost a 5 ppm range.

The first thing you should notice in the above image is that the highest CO2 readings are not over areas you would suspect. Instead the higher readings are over the Arctic, the Amazon, and parts of the Atlantic and Pacific Oceans. Much to the horror of the global warming alarmists, one of the lightest areas indicating low levels of CO2 is over the heavily industrialized Northeast USA. Granted this particular photo was March 2015, and for another month or year would be different. But the facts speak for themselves.

There is also the option to play the changing CO2 from 2002 to the present. If you position the Earth so that the Arctic and the Northern Hemisphere are seen and press play, you will notice over the subsequent 13 years how the image of increases in CO2 appear first in the Arctic and seem to migrate south towards the USA. You would expect the reverse to be true. But one animation is worth a thousand words. Here from NASA - home of the guys in GISS (Goddard institute for Space Studies, a laboratory in the Earth Sciences Division of NASA) who continually are cooking the books of the temperature record – is a simple animation of the historical CO2 changes since 2002. What these animations show is that CO2 comes from everywhere, except where you would think. After viewing these animations the only conclusion that can be drawn is that man has nothing to do with CO2 generation - NOTHING.

Below is another presentation of CO2 concentrations, this one from the Orbiting CO2 Observatory, OCO-2. Notice the dark spot over China reflecting an increase in CO2, which is no surprise with all their industrialization in the area. But look at that dark spots reflecting higher CO2 south of Greenland in the middle of the Atlantic Ocean, and south of the Bering Sea in the Pacific Ocean. Care to explain these higher concentrations of CO2 in the middle of the Atlantic or Pacific Oceans?

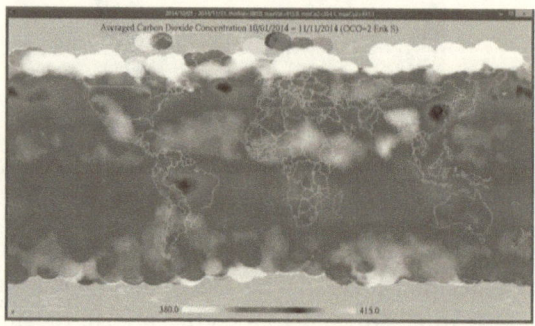

Fig 4g: Orbiting CO2 Observatory OCO-2
<http://eyes.jpl.nasa.gov/eyes-on-the-earth.html>

But do not fret. I am sure it will be only a short period of time before the scientists led by policymakers at NASA realize that these changes that happen frequently and often in various locations, for seemingly no reason at all, will have to be rectified, or should I say "homogenized", just like the homogenizing of temperature data is performed at NASA GISS.

Dr. Heinz Hug

All atmospheric scientists use 'established' data, regarding emission and absorption spectra of CO2. However, no one has ever questioned the accuracy of this data until now. Dr. Heinz Hug, not content with the 'established' data, conducted actual controlled laboratory experiments to confirm the 'transmission to extinction' properties of CO2 in air. What

Dr. Hug has found is nothing short of extraordinary - The 'established' data is an artifact that is wrong! (Hug, Dr. Heinz, "The Climate Catastrophe – a Spectroscopic Artifact", July 31, 1998)

Carbon dioxide *absorbs to extinction* all the longwave energy, characteristic of its wavelength spectra that can be absorbed. This means that while there are many miles of CO_2 above this point, there is no longwave energy remaining for it to absorb. If there was a doubling of CO_2 concentration, the distance required for the same *extinction* to occur would be reduced about 3%, or from about 40 feet to about 38.8 feet. But the same amount of heating from the energy absorbed by the CO_2 would remain the same. The only way for CO_2 or other greenhouse gases to absorb more energy is for more energy to be made available by the sun.

Hug concludes the radiative forcing for a CO_2 doubling must be reduced by a factor of 80! This means, the previous allocated 7.2 deg-C for the greenhouse effect cited by the IPCC (Krondratiev and Moskalenko, The Global Climate, Houghton) must be divided by 80 to obtain the true CO_2 doubling value of 0.09 deg-C.

This 0.09 deg-C increase is hardly measurable with anything except calibrated laboratory instruments. Then multiply this by our 3.27% contribution to the total and we arrive at the conclusion that man's contribution to CO_2 induced global warming is – drum roll please – 0.002943 deg C! Do you think any living thing will be aware or affected by this miniscule increase?

Summary

- CO_2 has increased about 85 ppm within the last 56 years.

- CO_2 was about 420 ppm in 1820 and again in 1942

- CO_2 concentrations follow a 230 year cycle

- CO_2 was 18 times higher over 500,000,000 years ago

- CO_2 increases occur about 800 years after the temperature increases.

Chapter 5: Temperature Record

Without hesitation, the most common expression describing the state of the recent couple of decades regarding the weather, or climate and global warming, is the temperature. Some people will allude to the heat waves over Moscow in 2010 or the US Midwest in July 2012 as proof of global warming. Any meteorologist without an agenda will tell us, these heat waves were caused by parked high pressure areas that after a few weeks moved on after the jet stream straightened out and released them.

On January 26, 1989, an article appeared in the New York Times entitled <u>U.S. Data Since 1895 Fail To Show Warming Trend</u>, and I quote " After examining climate data extending back nearly 100 years, a team of Government scientists has concluded that there has been no significant change in average temperatures or rainfall in the United States over that entire period." Reviewing data all the way back in 1989 NOAA scientists concluded there was no warming in the USA. So, what happened in the last 30 years that has suddenly transformed 'no warming' into climate change, climate disruption, climate catastrophe, etc.?

Global Temperature Record Histories

There have been many stories concerning supposed global warming and climate change. They all have supposed roots in what is known as the 'temperature record'. There are three main global temperature record histories: The NOAA record, the NASA GISS (GISSTEMP) record and combined CRU-Hadley (HadCRU) record. All three global averages are based on the same underlying archive, the Global Historical Climatology Network or GHCN (McKitrick, Ross, A Critical Review of Global Surface Temperature Data Products, July 26, 2010). The GHCN stations consist of 6,000 temperature, 7,500 precipitation, and 2,000 pressure measuring stations spread around the world.

Temperature Measurement Station Designs

So how are temperatures measured? On land, they are customarily measured by instrumentation housed in Stevenson Screens.

Fig 5a: Stevenson Screen with Access Door Opened
<http://www.bom.gov.au/climate/data/acorn-sat/documents/ACORN-SAT_Observation_practices_WEB.pdf>

These classical surface temperature measurement stations consist of louvered, ventilated, wood structures on a legged enclosure, containing instruments and/or sensors about 5 feet above grade. These enclosures are usually painted with white latex paint to provide a high degree of reflectivity. As these enclosures age there is a buildup of dust and dirt, and/or paint might start pealing off, and as a result their reflectivity will decrease. Over a long period of time as these stations age, their readings will be skewed to the warming side, even when there may be no warming at all outside. In theory, if these stations were periodically repainted, these artificial temperature increases would probably not occur. As with just about everything else these days, maintenance is not usually a high priority item. For the global warming/climate change crowd, why should there be any incentive to correct a condition, which might show that there is no warming at all?

Since 2005, NOAA has overseen the installation of 114 spanking-new, state-of-the-art temperature measurement stations throughout the lower 48 States. The locations of these new stations were specifically selected to be far away from urban developed areas to avoid the 'heat island effects' of these compromised locations. There has not been too much news about these new stations and for good reason. The measurements from these stations reveal that since 2005, there has been no temperature increase at all. You can be sure, if there were temperature increases, the news would be splattered all over the front pages.

Regardless of the design of these temperature measurement stations, they are all measuring temperatures only five feet off of the Earth's surface. But what changes are occurring above the surface? From a practical point of view, 100% of the atmosphere is contained within the first 30 miles (50 kilometers). So, what's going on way up high?

To Surface or Not to Surface

The entire scare of global warming is based on the claim of increasing Earth SURFACE temperatures. When global warming alarmists refer to temperatures, they are ALWAYS referring to SURFACE temperatures, not the overall temperature of the entire atmosphere. The penchant for using only surface temperature measurements is done on purpose for numerous reasons. First, since the air temperatures above the surface or lower stratosphere are cooling, they are discounted. Second, even with no further increase in fossil fuel burning or CO_2 increase, surface temperature measurements over time increase due to encroachment of civilization on measuring instrumentation. This is known as the Urban Heat Island (UHI) Effect. Third, although recently 75% of temperature measuring stations have been closed, more stations have closed in remote (cooler) locations than more conveniently accessible, local (warmer) locations. Fourth, since there are no stations above latitude 80, the temperature of the Arctic is a complete guess, even though half century long measurements have been made by the Danish Meteorological Institute which show a cooling Arctic. But, alas, these DMI measurements are not considered as part of the official temperature record. If they showed a warming trend, do you think they would receive acceptance? You Betcha.

It is a fact that the surface temperature record according to measurements taken at surface stations is increasing. This is depicted in the graph below. The slope of the readings works out to be about 1.0C within 50 years, or about 0.02C/year, positive. No argument here that surface readings are increasing, but for the reasons cited above.

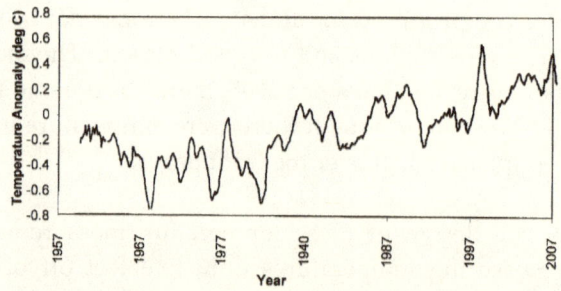

Fig 5b: Global Surface Temperature Anomalies
<http://hadobs.metoffice.com>

However, looking at the temperature measurements above the surface and below 3-km (1.86 miles) the global lower stratosphere temperature anomalies are decreasing as shown in the figure below. The blips marked 1, 2 and 3, represent spikes in temperatures caused by volcanic eruptions of Agung, El Chichon, and Mt. Pinatubo. The change in temperature works out to be about 2.0C in 50 years, or 0.04C/year, negative.

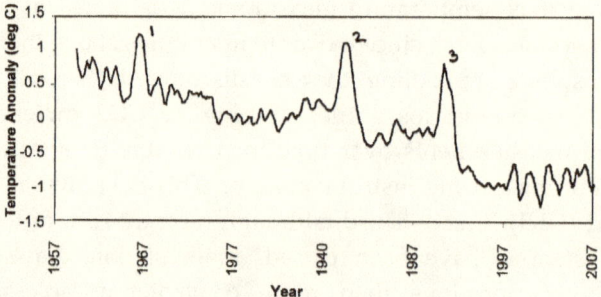

Fig: 5c: Global Lower Stratosphere (0-3km) Anomalies
http://hadobs.metoffice.com

52

What, if we combined the two above temperature anomalies. Would they cancel each other out? Graphically, at first glance, you would think so, since the density of the atmosphere changes with altitude for the first few km is relatively linear. Therefore, using a few proportions we can arrive at a reasonably close estimate. Taking the lower stratosphere average of 0-3-km as being 1,500 meters and weighing this against a surface thickness of say 10-meters, it is obvious the surface temperature influence is about 10/1,500 or 1/150th of the global lower stratosphere. This is a reasonable conclusion based on the assumption that all of the infrared energy heating due to CO2 is absorbed to extinction within the first 10-meters above the Earth's surface. So, the resultant slope of the combined two temperature anomalies works out to be:

$$-0.04 + \{ [(0.02 - (- 0.04)] / 150 \} = - 0.0396C/year.$$

In other words, the global <u>surface</u> temperature anomaly almost always measured five feet above grade doesn't represent anything. Within the first 1,500 meters above the surface of the Earth for the last 50 years the Earth has been cooling at the rate of 0.0396C/yr.

<u>Temperature Measurement Station Geographical Locations</u>

Below is a map depicting GHCN stations as they exist today. Notice the heavy concentration of stations in the United States. We have 1251 or about 17% of the world's 7,364 stations. Yet the lower 48 States has an area of only 1.58% of the Earth's surface.

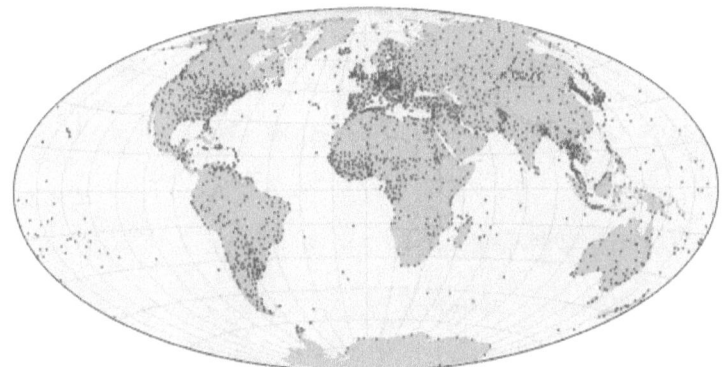

Fig 5d: Global Temperature Measuring Station Locations
<http://www.earthobservatory.nasa.gov>

The Antarctic is another quandary. Of the two dozen or so stations in Antarctica, except for two, all of them are located on the coastline. More significantly, about half a dozen are located on the peninsula, where in addition to being closer to warmer ocean temperatures and farther away from polar conditions, there is the further influence of heat released from underwater volcanoes in the vacinity. The coldest point in Antarctica is the South Pole. Yet, amazingly, there are no stations shown at the South Pole. We know of course there must be stations there. The problem is they keep getting buried under snow and ice.

Then there's Greenland. One of the 'canaries in the coal mine' that seems to be the focus of melting and whose ice cap is always threatened with sliding into the sea raising sea levels by who knows what. There are only seven stations there. Not surprisingly, they are all located on the coastline. None are located inland. I don't know about you, but there is a big difference between measuring temperature at the shore and 50 miles inland. If you don't appreciate this, try comparing summertime temperatures in Cocoa Beach with those of Orlando in Florida; a comfortable sea breeze versus a stagnate microwave oven.

Do you think this concentration of stations within the largest industrialized nation is being properly accounted for with the appropriate compensation? How can we be sure?

The Temperature Keepers

So, exactly, where are these surface temperature measurements being taken and who is performing the measurements? The most recently accepted data set is the GHCN Version 2 temperature database. It is derived from 31 data sets from around the world. (Peterson and Vose, An Overview of the Global Historical Climatology Network Temperature Database, Bulletin of the AMS, Vol. 78, No. 12, December 1997)

US Temperature Measurement

The Climate Reference Network (CRN) Rating Guide - adopted from NCDC Climate Reference Network Handbook, 2002, has specifications for siting (section 2.2.1) of NOAA's new Climate Reference Network. Of the approximately 6,000 GHCN stations worldwide, there are 1251 GHCN stations in the United States. They are classified by Class Rating Number ranging from CRN=1 (Best) to CRN=5 (Worst).
Class 1 (CRN1) - Flat and horizontal ground surrounded by a clear surface with a slope below 1/3 (<19deg). Grass/low vegetation ground

cover <10 centimeters high. Sensors located at least 100 meters from artificial heating or reflecting surfaces, such as buildings, concrete surfaces, and parking lots. Far from large bodies of water, except if it is representative of the area, and then located at least 100 meters away. No shading when the sun elevation >3 degrees.

Class 2 (CRN2) - Same as Class 1 with the following differences. Surrounding Vegetation <25 centimeters. No artificial heating sources within 30m. No shading for a sun elevation > 5C.

Class 3 (CRN3) (error >=1C) - Same as Class 2, except no artificial heating sources within 10 meters.

Class 4 (CRN4) (error >= 2C) - Artificial heating sources <10 meters.

Class 5 (CRN5) (error >= 5C) - Temperature sensor located next to/above an artificial heating source, such a building, roof top, parking lot, or concrete surface."

The following is a summary of the above site quality:

CRN	Quality	Temperature Error	Perecent of Total
1	Best	< 1C	3
2	Good	< 1C	8
3	Fair	>= 1C	20
4	Poor	>= 2C	58
5	Worst	>= 5C	11

As you can see from the CRN stations below in Fig 5e, 69% of the 1251 stations fall into the poor or worst classification. This means that there is a temperature error between 2C to 5C, or more than 2/3's of the stations have an error of about 3.5C. So, how will this average station error of 3.5C detect the 1.5C increase of burning all fossil fuels? For that matter, how will these stations detect any smaller temperature increases at all?

The atmosphere contains about 750 gigatons of carbon. If the entire world's known reserves of 4,000 gT (as of 2010) of fossil fuels were burned overnight, the CO_2 concentration of the atmosphere would increase from 390-ppm to 2,450-ppm the next day. This would result in a temperature increase of 1.764C (see Figure 6b, Point I explanation). So, with 69% of the CRN stations classified in the "poor" or "worst" category with a temperature measurement error of at least 2C or more than 5C, just how much confidence can you have in detecting a 1.5C increase, let alone a 0.0396C/year decrease?

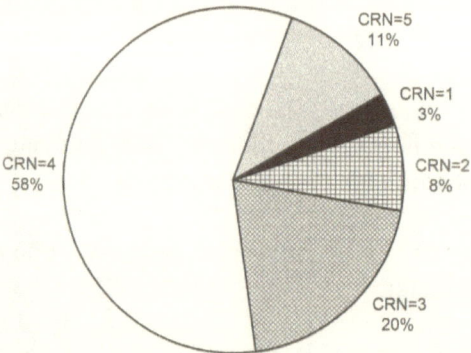

Fig 5e: Surveyed CRN Site Quality Rating

Temperature Measurement Station Site Conditions

Besides the condition of the weather stations themselves, there is the issue of where these stations are situated. The condition of many of these sites has much to be desired. Some are appalling. Some are hilarious. What follows is one example of a photo taken documenting the condition of a temperature measurement site.

Fig 5f: Detroit Lakes MN GHCN Site
<http://www.norcalblogs.com/watts/2007/07/
how_not_to_measure_temperature_23.html>

The site arrangement at Detroit Lakes MN is hysterical. The classical wood-louvered Stevenson screen and a companion Automatic Weather

Station (AWS) cylindrical Stevenson screen are just behind the railing. It is not known if the instruments in both these screens are active. The louvered screen is in obvious need of refurbishment and a new coat of white Latex paint. Otherwise, over the years the temperatures recorded inside will get warmer and warmer, just because more heat is being absorbed by the screen outside, which is getting dirtier and dirtier. It wouldn't hurt to re-level this instrument enclosure, either. In the foreground are two air conditioning condensing units just inches away from these screens. Do you think this is a good location for taking the temperature of the atmosphere at Detroit Lakes?

Up until the year 2000, these condensing units were located on the roof of the building off to the right of the picture. After 2000, they were relocated to the ground as shown in the picture above. The reason for the relocation is unknown. A spike in temperature measurement can be seen in the graph below. The Annual Mean Temperature jumped in 2000 from about 3.2C to 6.2C, or a 3C step.

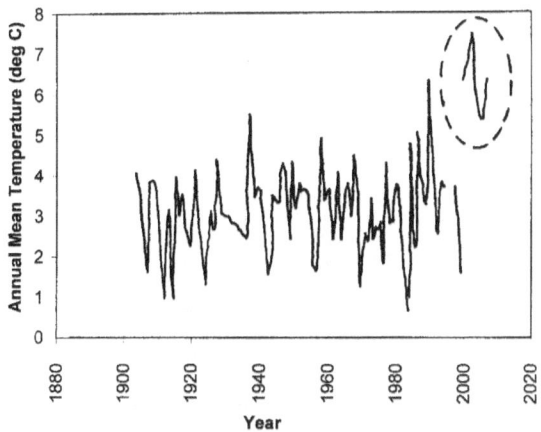

Fig 5g: Detroit Lakes Temperature Record 1980-2006

It doesn't take a genius to figure out that the accuracy in temperature measurement for Detroit Lakes is called into question, if two heat producing condensing units are spewing extra heat all around.

Detroit Lakes MN is not the only station with severely compromised temperature measurement arrangements. A collection of photos taken of other poorly situated GHCN sites can be found in the report Is The Surface Temperature Record Reliable, available online at http://watts upwiththat.files.wordpress.com/2009/05/surfacestationsreport_spring09.

In this report you will see photos of weather stations next to asphalt parking lots, on blacktarred building rooftops, near large utility transformers, on city streets near sidewalks, near air conditioning units, adjacent to nice warm clarifiers inside wastewater treatment plants, etc. Nearly 9 out of 10 stations fail to meet the National Weather Service (NWS) criteria of having to be more than 100 meters away from any artificial heating or reflective surface <www.surfacestations.org>. The dozens and dozens of photos are testament to the deplorable condition of the entire temperature measurement system.

This begs two questions: First, who is in charge of locating these weather enclosures and their instruments and approving their locations? Second, who is in charge of collecting, processing and analyzing these weather instruments? In the US, the task of analyzing instrument data up until 2014 had been headed up by Dr. James E. Hansen.

Dr. James E. Hansen

Who is James E. Hansen? He was born in Iowa in 1941 and received his BS, MS and PhD from the University of Iowa in 1963, 1965 and 1967, respectively. Since then, he has spent most of his entire career grazing in the public trough at NASA GISS.

Fig 5h: Dr. James E. Hansen – NASA GISS
< NASA Photo>

The picture above shows Dr. Hansen testifying before Congress on June 23, 1988 about global warming. If it wasn't for the name plate in front of him, you might confuse him with Agent Smith in The Matrix. I can hear him now: "What good would a phone be, Mr. Anderson, if you cannot speak?" It is not clear when Dr. Hansen had his epiphany on climate change from global cooling to global warming between 1971 and 1988.

But I suspect the trough of bigger public money to study global warming had something to do with it.

Hansen's first notoriety occurred while a research assistant to Dr. Ichtiaque S. Rasool at Columbia University. Dr. Hansen was responsible for the development of computer programming, which permitted Dr. Rasool to forecast an impending catastrophic change to the climate from dust released by the burning of fossil fuels. (Washington Post, July 9, 1971, "U.S. Scientist Sees New Ice Age Coming") And what did this computer program forecast? The next impending Ice Age! That's right, folks. Dr Hansen back in 1971 was involved with the production of a study predicting global cooling resulting from the burning of fossil fuels.

So, what is Dr. Hansen doing today? He is running around the country extolling the horrors of global warming from fossil fuel burning. He is proclaiming coal fired power plants as "factories of death".

So, how does a government employee - a well paid and highly influential government employee - continue to get compensated from his government job while attending his outside political interests? Good question. More importantly, how does Dr. Hansen maintain any credibility as an impartial custodian of the temperature record?

What makes matters worse, is how much and from where Dr. Hansen gets additional monies supplementing his government salary. Back in 2006, Dr. Hansen received $720,000 from a foundation run by George Soros <http://newsbusters.org/blogs/noel-sheppard/2007/09/26/nasa-s-hansen-mentioned-soros-foundations-annual-report>. But the stench gets worse.

George Soros

Who is George Soros? George Soros was the world's 15th wealthiest man according to a 2012 Forbes survey. He was co-founder of the Quantum Fund, one of the top performing hedge funds of all time. He is a backer of liberal causes, shadow governments, and is an atheist with no conscience, whatsoever. In a December 20, 1998 interview with Steve Kroft on 60 Minutes, George Soros described his happiest year:

> KROFT: My understanding is that you went out with this protector of yours who swore that you were his adopted godson.

SOROS: Yes. Yes.

KROFT: Went out, in fact, and helped in the confiscation of property from your fellow Jews, friends and neighbors.

SOROS: Yes. That's right. Yes.

KROFT: I mean, that sounds like an experience that would send lots of people to the psychiatric couch for many, many, years. Was it difficult?

SOROS: No, not at all. Not at all, I rather enjoyed it.

KROFT: No feelings of guilt?

SOROS: No, only feelings of absolute power.

George Soros as a teenage Nazi sympathizer had no qualms about stealing the property of half a million of his fellow Jews that eventually were exterminated, and Dr. Hansen had no qualms about using that $720,000 from George Soros to continue his work on the global warming scam.

USA Temperature Homogenizing

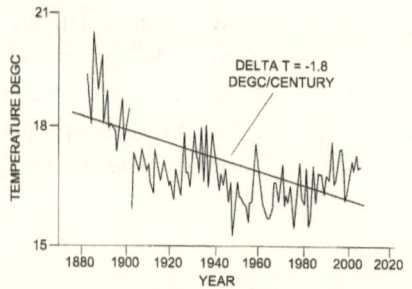

Fig 5i: Orland CA Temperature History as of 2007
<http://pracphilosblog.files.wordpress.com/
2010/02/orland_station_plot_20071.gif>

The above graph is an example of how Dr. James Hansen and his "Band of Homogenizers" rework temperature data history. In 2007, if you went to the NASA GISS web site, you would find the above annual mean temperature history from 1880 to 2007 for Orland CA. The temperature trend works out graphically to be about a 1.8C/Century decrease. No global warming here.

60

But, if you went to the GISS web site in 2010, you will find the following "adjusted" temperature history. Without any explanation as to why the data from 1880 to 1900 disappeared, the result of a 1.8C/Century decrease in 2007 has been suddenly transformed into a 1.5C/Century increase in 2010. Voila! Instant global warming!

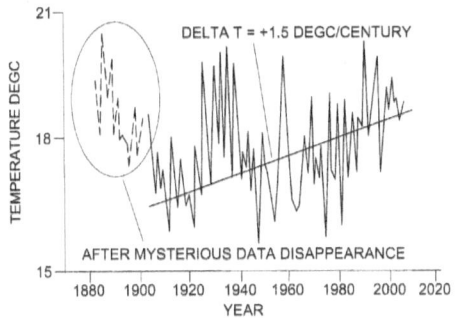

Fig 5j: Adjusted Orland CA Temperature History as of 2010
<http://pracphilosblog.files.wordpress.com/
2010/02/orland_giss_station_plot_raw_121109.gif>

What the reader should find disturbing is the arbitrary deletion of the first 20 years of temperature history from 1880 to 1900. Perhaps this adjustment was, indeed, necessary to make this data consistent with other data with shorter time frames. Perhaps not. But, if it was the intent to show an increasing trend in temperature, then the temperature data before 1900 just had to go. And that is likely what NASA GISS and Dr. James Hansen have done.

Thus, with the wave of a magic wand, Dr. James "thumb on the temperature scale" Hansen (from here on I will use "Totts" as Dr. James Hansen's middle name), has transformed a clear 1.8C/Century temperature decrease into a 1.5C/Century temperature increase. I am also offended how the word "raw" got introduced into the revised plot's website reference, giving the impression that the revised plot was based on "raw" data, not "revised" data.

One of the reasons you are hearing that every year is the warmest year on record is the constant tinkering of temperature data by NOAA and NASA GISS. They continue relentlessly homogenizing data, deleting old warmer data to give the appearance of warming trends when if the data were left alone would indicate a cooling trend, using an ever decreasing number of real stations and replacing them with an ever increasing number of fake stations, using measurements with ever decreasing quality due to societal encroachment, or readings taken from improperly maintained weather stations. From my perspective, the scientific fraud that is going on at NOAA and NASA GISS regarding temperature measurement and their reporting is criminal and warrants a Congressional investigation.

Worldwide Temperature Measurement

Homogenizing and adjustments, besides being severe and arbitrary, happen outside the USA, as well. Take the graph below at Darwin Airport in Australia, for example.

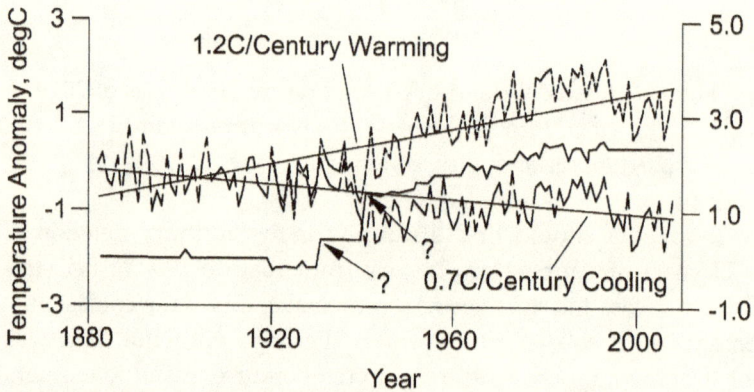

Fig 5k: Raw & Adjusted Temperatures Darwin Airport, Australia
<http://hotair.com/archives/2009/12/09/east-anglia-homogenization-falsified-declines-into-increases/>

The "adjusted" data shows a warming curve of 1.2C per century. The raw data curve shows a cooling trend of 0.7C per century. Notice the step increase in adjusted temperature (indicated with a "?") around 1930 and around 1940. These step changes could have been the result of relocating instrumentation to a different site, changing of instruments, a new parking lot or structure being built nearby, etc.

Worldwide, the historical temperature record doesn't look much better. The worldwide GHCN version 2 temperature database is derived from 31 datasets with a total of 24,365 stations. Yet, this century-scale set of observations is from only about 7,000 stations <http://www.ncdc.noaa.gov/oa/climate/ghcn-monthly/images/ghcn_temp_overview.pdf>. So, what happened to the data from the other 17,365 stations? Did this data get tossed into the trash can, because it did not fit?

In 1996 the number of stations plummeted from about 6,000 to 1,500. I do not know if this was a result of down-sizing due to economic reasons, or if the results from the closed stations would have resulted in a temperature record that was cooler and therefore unwanted. As bad as this reduction in sample size is, it is magnified by the sample becoming increasingly skewed towards stations located at airports, migrating from colder to warmer latitudes, and migrating from higher (cooler) to lower (warmer) altitudes. <http://rossmckitrick.weebly.com/uploads/4/8/0/8/4808045/surfacetempreview.v3.pdf>. All of this station sample migration skews the resultant world temperature higher, regardless of whether temperatures increase, or even decrease.

A reduction in the number of temperature measurement stations from 24,365 to 1,500 is appalling. Thus, the impact of a temperature change in one station today has more than 16 times the impact of a temperature station back when there were 24,365. stations. So, it should not be a surprise to realize that any adjustment in the temperature record of 100 years would have an amplifying effect today because of this large station reduction.

Real Data versus Fake Data

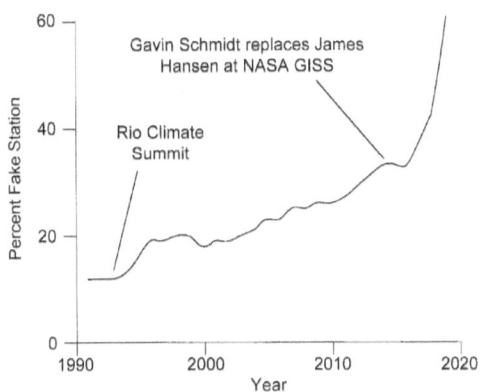

Fig 5l: Percentage of USHCN Stations With Fabricated Data
<https://realclimatescience.com/wp-content/uploads/2019/02/PercentOfUSHCNMonthlyTemperatureDataWhichIsFabricated_shadow.jpg>

Most people would be under the impression that data from temperature measuring stations are recorded by temperature measuring instruments at real temperature measuring stations. But you would be very wrong.

For the last 20 years, more and more temperature data is being simply fabricated out of thin air. Below is a graph of the percentage of USHCN stations in which data is fabricated.

Up until 1993, only about 12% of station temperatures were fabricated. This rather flat trend abruptly changed around 1994. Perhaps inspired by the United Nations Conference on Environment and Development (UNCED), Rio de Janeiro, 3-14 June 1992 (the first 'Earth Summit'), the percentage of fabricated USHCN Station data began a steady increase. In 2014 Gavin Schmidt took over the reigns from James Totts Hansen. Not too soon after that the percentage of fake stations took off like a rocket ship. The percentage of fake stations in 2016 was 33%, in 2017 was 38%, in 2018 was 44%, and 2019 was 61%. That's not a misprint. Between 2018 and 2019, the percentage of fake stations increased from 44% to 61% in one year. At this rate the percentage of fake stations would hit 100% by 2023!

Can you imagine the opportunity this makes available to generate - what otherwise would be no temperature increase at all with real data - a temperature increase of crisis proportions with fabricated data? Do you think Dr. James Totts Hansen or his successor Dr. Gavin Schmidt had anything to do with this?

With all the talk of the Artic Ice cap disappearing, you would think a reliable temperature record would be an absolute necessity in this frozen part of the world. But, there are no GHCN stations at the North Pole, nor for that matter anywhere north of latitude 80N. To be fair, there is little land in this area; only water and ice. That hasn't stopped the Danish Meteorological Institute (DMI) which has been faithfully recording temperatures in the Arctic since 1958, the same International Geophysical Year that CO2 measurements started on the Mauna Loa volcano in Hawaii. One has to wonder why Dr. James Totts Hansen has not spliced this crucial Danish data set into the database to fill this most important void. The answer is very simple. The DMI records show that during the last 52 years, there has been no warming of the Arctic at all. You can be sure, if the DMI record showed warming, James Totts Hansen would have immediately jumped at the opportunity of splicing this data set it into his master homogenized database.
This whole adjustment and homogenization process involving temperature measurement smells. And the stink is coming from NASA GISS, from Dr. James Totts Hansen, and now his successor Dr. Gavin

Schmidt.   The law of averages would suggest that half of any temperature adjustments would result in temperature increases and half the adjustments would result in temperature decreases. I suspect that the temperature increase adjustments completely overwhelm the temperature decrease adjustments (if there are any), and there is no good explanation for this, other than global warming agenda bias of NASA GISS.

To be fair, over the last 100 years inconsistencies have evolved for the temperature stations in the network.   These inconsistencies include station openings, station closings, station location changes, splicing together new data sets to existing data sets, instrument replacements, times of observation, methods used to calculate readings, and changes to station environments.   To bring all this inconsistent data together requires a process of "homogenizing". The question is: who is doing the homogenizing and how faithfully are they executing changes without introducing any agenda biases?

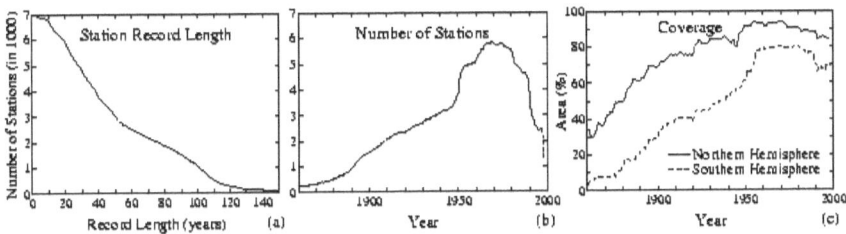

Fig 5m: Gistemp Station Data
<http://data.giss.nasa.gov/gistemp/station_data/>

At a glance you can see the problems being encountered trying to bring GHCN data together in a coherent format. Graph (a) above shows there are about only 1,000 stations that have a record length of 100 years, and 6,000 stations which have a record length of only 20 years.   Graph (b) shows the number of stations increasing from about 250 in 1860 to almost 6,000 in 1970. However, since then about 81% of the stations have closed leaving 1,113 as of 2010.   These closings have increased the number of airport stations from 30% to 50%. International airports have huge areas of concrete some 18" thick which absorb heat from the day and are very reluctant to release this inventory of heat after sundown. So, like the inner cities, while the suburbs are cooing off at might, airports are still boiling. Graph (c) shows the difference in coverage between southern and northern hemispheres.

Perhaps the cruelest hoax of all, and the basis for all the Global Climate Models (GCMs), is the surface temperature record of the world. This record is constantly revised and "adjusted" by an elite group of individuals at only three agencies. The true reason is that the GCMs still cannot be relied upon for accurately projecting future climate. How can they? They cannot even forecast the weather a few days in advance. How can they predict climate decades or even a century in advance? So, rather than adjust the models which keep yielding incorrect results, "the elite" at "the three" keep incessantly fiddling with the data hoping they can make the models work. The sad fact of the matter is that many of these stations measuring temperatures are dialed into climate models, that are used to project future temperature trends, that are used to sway public policy, that are used to pass legislation on global warming. The entire temperature record established to date by the GHCN is a corrupted pile of rubbish.

It doesn't make things easier when the people in charge of documenting the temperature throw away the data they are entrusted to secure. Incredibly this is what has happened. "Scientists at the University of East Anglia (UEA) have admitted throwing away much of the raw temperature data on which their predictions of global warming are based". The CRU in a statement on its web site said: "We do not hold the original raw data but only the *value-added* (quality controlled and homogenized) data". (Leake, Jonathan, Climate Change Data Dumped, The Sunday Times, November 29, 2009). This means you will just have to trust the adjusted "value-added" data which has been maintained. This means there is no method to determine how the raw data was "homogenized". This means the "quality controlled" data which remains cannot be back-checked, since there is nothing to compare it to. But, isn't that what science is all about; the ability of other scientists to use the same data and independently duplicate results? Apparently, not as far as CRU is concerned.

The CRU's reason for throwing out the old paper and magnetic data was: they would not have room for it in their new building. To this I say "You have to be kidding". One would think, if the CRU was moving to a new building that the new building would not only be more modern, but also more spacious. I cannot envision CRU moving to an older, smaller building. Logically, there had to be enough room for this data. What else was taking up so much space that raw data had to be discarded? And even if there wasn't enough physical space, did CRU ever hear of flash drives, hard drives, off-site data storage, "the cloud", etc.? This explanation for throwing away data is vacuous, inexcusable, and way too convenient. No, I'll go further to say the only conclusion that can be drawn is this: the whole surface temperature measurement game is beyond suspicious. It is corrupt.

The Temperature Record for Last 600,000,000 Years

The graph below represents 600,000,000 years of proxy temperature data compiled by Dr. Christopher R. Scotese in his PALEOTEMP Project.

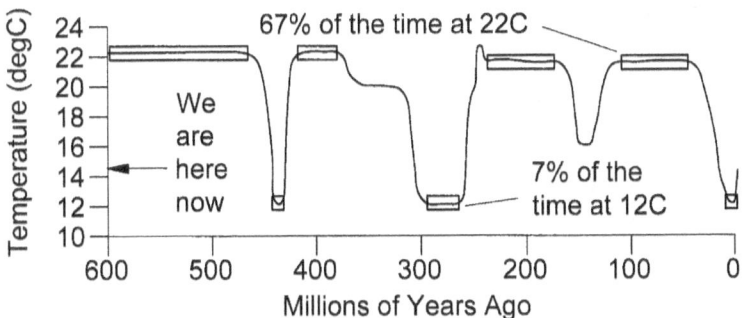

Fig 5n: Temperature for last 600,000,000 Years
<www.scotese.com>

Rather than paraphrase what is going on with the above curve, I quote a talk given by Astrophysicist Edwin X. Berry PhD on March 7, 2014, which also appeared in the <u>Daily Inter Lake Op Ed</u> of August 31, 2014 <https://edberry.com/blog/climate/climate-politics/global-warming-eco-religion-2/: >

'In the last 600 million years, our solar system passed through four spiral arms of our Milky Way. Earth's average temperature was 22C when outside our galaxy's spirals arms but decreased when inside each spiral arm, where cosmic rays are stronger.

From 450 to 420 million years ago, Earth passed through the Perseus spiral arm and average temperature dropped to 12C even while CO_2 concentration was at 4,500 ppm, 11 times today's CO_2 concentration. These temperature drops during high CO_2 contradict the global warming hypothesis. Upon exiting the Perseus spiral arm, Earth's temperature returned to 22C.

From 320 to 270 million years ago, Earth passed through the Norma spiral arm and the temperature again dropped to 12C, rising again to 22C when Earth exited this spiral arm.

From 150 to 130 million years ago, the earth passed through the Crux-Scutum spiral arm and the temperature dropped to 16C at the end of the 100 million year Triassic-Jurassic age,

For the last 20 million years, Earth has been in the Orion Spur spiral arm and temperature dropped to 12C to 16C.

This shows CO2 has little effect on Earth's temperature but cosmic rays cause temperature to drop by increasing cloud cover.'

Notice above that about 67% of the time temperatures we're at about 22C, and only about 7% of the time temperatures we're at about 12C. Historically, we have spent more than 9 times the amount of time near the 22C high than at the 12C low. Right now the world temperature is about 14.5C or about 25% off the bottom of the historical low. So, what's the problem?

Hottest Year Ever?

It seems recently that with each passing year, we are being told 'this year is the hottest on record'. These claims are based on the temperature records of the last hundred years or so being homogenized, twisted, corrupted, and faked by the scientists at NASA GISS. However, when looking back 10,000 years covering the present Interglacial Warmup Period (IWP), a totally different picture comes into view.

The graph above plots 10,000 years of ice core temperature data compiled by Dr. Richard B. Alley during the second Greenland Ice Sheet Project, GISP2. These ice core temperatures are indicative of the surface air temperatures that existed when these layers of ice were formed.

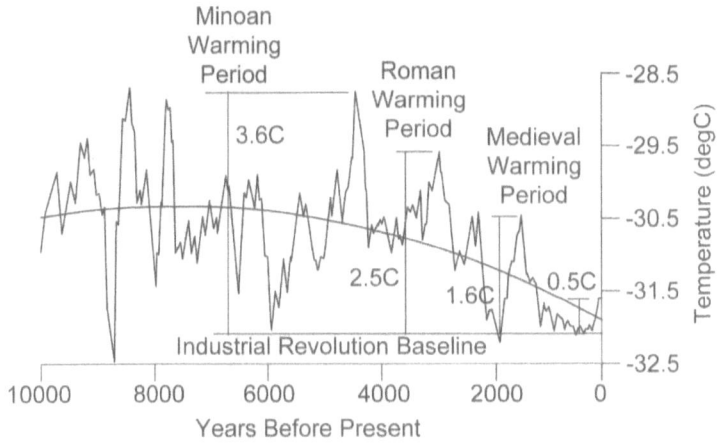

Fig 5o: Temperature for last 10,000 Years
(Alley R. B., The Younger Dryas cold interval as viewed from central
Greenland, Journal of Quaternary Science Reviews, 19:213-226)

Notice in that graph that we are about 0.5C warmer than the baseline temperature of the Industrial Revolution. The IR was the period where the burning of fossil fuels started to escalate around 1760. If you go back to the Medieval Warming Period about 1,500 years ago, temperatures were about 1.6C higher than the IR baseline. Going back further to the Roman Warming Period about 3,200 years ago, temperatures were about 2.5C warmer than the IR baseline. Finally, we see the Minoan Warming Period about 4,300 years ago where temperatures reached about 3.6C higher than the IR baseline. It is plain to see, the warmth experienced thousands of years ago were many times higher than the recent paltry increase since the Industrial Revolution.

No wonder the Anthropogenic Global Warming (AGW) worshipers are alarmed at this evidence. Look at the trend line. For the last 6,000 years the trend is down - and accelerating downward. So much for the "warmup" in this IWP. Oh, and by the way, Dr. Alley has been a contributor to past IPCC Assessment Reports.

Summary

- Of the 1,500 GHCN stations in the USA, 89% are classified as "fair", "poor" or "worst", having accuracies of greater than 1, 2, and 5C, respectively. Only 8% are rated as "good", and only 3% are rated "best", both with an accuracy of less than 1%.

- As society grows and expands it begins to encroach upon these stations. This causes the appearance of ever increasing temperature measurements when in fact there is no increase in temperature at all.

- The number of faked stations has increased from around 3% between 1950 and 1992 to 61% in 2019; a 2,000% increase in faked data over the last 26 years.

- There are no temperature measurement stations north of Latitude N80.

- Of the 12 stations in Antarctica, only 2 are inland, the others are on the coast.

- Of Greenland's 7 stations, all are on the coast, none are inland.

- For 52 years the Danish Meteorological Institute has taken temperature measurements in the Arctic, but their data has never been assimilated into the other data sets.

- Of the 114 new real temperature measurement stations installed since 2005, all of them have recorded no increase in temperature at all.

- Right now the world temperature is about 14.5C or about 25% off the bottom of the historical low.

- Within the last 10,000 years there have been three warming periods many times higher than the one experienced since the Industrial Revolution.

"All truth passes through
three stages. First, it
is ridiculed. Second, it
is violently opposed.
Third, it is accepted as
being self-evident."

– Arthur Schopenhauer

Chapter 6:
Paired Records

Temperature records and CO2 records are perhaps the most studied, but there are other records also deserving of attention. Many of these are paired records.

Temperature Versus CO2 Changes - Linear

The circled area in Figure 6a below represents the data taken within the last 50 years at Mauna Loa. On first inspection, it would appear reasonable to project what the temperature would be at higher CO2 levels by extending a tangent line from that portion of the curve. However, it is apparent that projecting this tangent line would yield incorrect results, because further CO2 increases yield smaller and smaller increases in temperature.

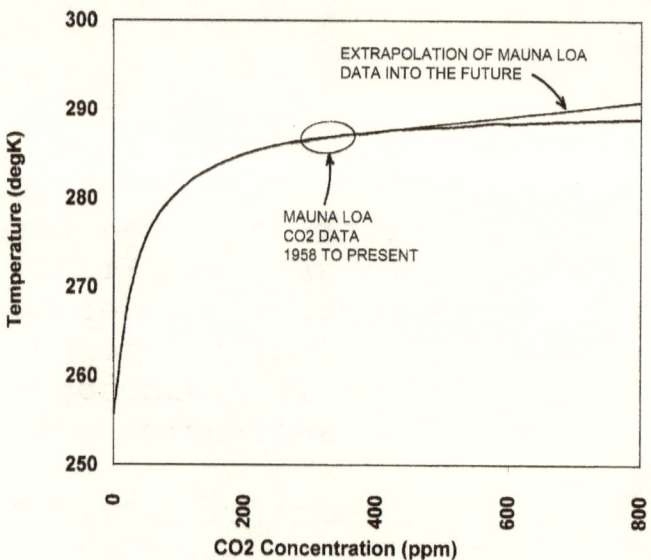

Fig 6a: CO2 versus Temperature, Linear
<www.brnurosci.org/co2.html>

To see what the value of temperature would be at 1 million ppm in the above graph you would need a sheet of paper that was 281 feet wide, or almost a football field in length. Yet, without a graph this wide, it is obvious that the curve above is approaching a horizontal line and that temperature increases beyond 800 ppm are going to be smaller and smaller. So, how do we get to see what happens beyond 800 ppm more accurately? Look at the same data, semi-logarithmically.

Temperature Versus CO2 Changes – Semi-Logarithmic

Below is the CO2 versus Temperature plot from 1 ppm to 1 million ppm CO2 shown in semi-logarithmic fashion. The beauty of this plot is its simplicity. The curve is an absolutely straight line. It is based on the formula:

$$K = 255 + 4.85 * \ln(x)$$

where: K = the temperature in degrees Kelvin
ln = the natural logarithm, and
x = the CO2 concentration in ppm

This formula was developed using the data from the source of the previous Fig 6a (www.brnurosci.org/co2.html).

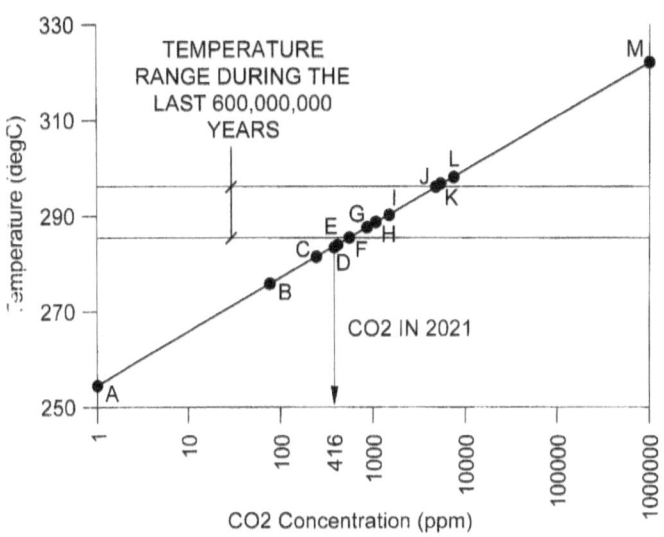

Fig 6b: CO2 versus Temperature, Semi-Logarithmic

What is significant about this semi-log plot (x [bottom] axis values are logarithmic, y-axis [left] values are linear) is that it shows for every ten-fold increase in CO2, there is the same 11.167C increase in temperature. Thus, the increase from 1 to 10 ppm CO2 results in the same temperature increase as from 100 to 1,000, or 1,000 to 10,000. Let's look at some of the important points on the above chart.

Point A shows the line intercept temperature of 255K at 1 ppm.

Point B shows the minimum CO2 concentration necessary for photo-synthesis at about 90 ppm. (D. N. Moss, The Limiting Carbon Dioxide Concentration for Photosynthesis, Nature **193**, 587, 10 Feb 1962). That's right folks, below 90 ppm, photosynthesis virtually stops and the green world we all know would cease to exist.

Point C shows where the CO2 levels are reached inside a greenhouse by the end of the day, about 300 ppm.

Point D shows the CO2 concentration of the atmosphere as of 2021 at 416 ppm, with a corresponding temperature of 11.1C.

Point E shows the levels reached just before sunrise in a greenhouse, about 450 ppm, with an equivalent temperature of 11.48C.

73

Point F indicates the CO_2 concentration of 501 ppm resulting with the 600,000,000 year historical average minimum temperature of 12C.

Point G shows what the ppm *should* be at the 2021 average world temperature of 14.936C at 917.7 ppm. But the corresponding temperature for the CO_2 concentration in 2021 (Point D) *was* 416 ppm. This disconnect indicates – and rightly so – that other factors are entering into the equation besides CO_2 concentration.

Point H shows the concentration of 1,000 ppm which is the ideal concentration of CO_2 required for optimum growth in a greenhouse. In fact, some greenhouse operators directly inject CO_2 into their enclosures and/or use special space heaters which vent their combustion products inside the greenhouse to keep CO_2 at elevated, beneficial levels. Plants just love living in an environment of elevated CO_2. As far as the Supreme Court and the EPA is concerned, CO_2 is a pollutant. As far as plants are concerned, high CO_2 concentrations are nirvana.

Point I shows what the CO_2 level would be, if we burned every last bit of the known 5,137 Gt of fossil fuel reserves on Earth. (Note: this total was calculated by the author from various on-line sources). The atmosphere already contains 2,362 Gt of CO_2 and after absorbing another 5,137 Gt would increase to 7,499 Gt. This would result in the CO_2 concentration increasing from the present 416 ppm to 1,319 ppm, and a resulting temperature of 16.7C. In other words, the increase in temperature if we burned every pound of fossil fuel reserves overnight would be 16.7C minus Point G's 14.936C, or 1.764C. How can this be? The temperature increase of the Medieval Warming Period was 1.6C, Roman Warming Period was 2.5C, and Minoan Warming Period was 3.6C (Fig 5o). And back then, fossil fuel burning was nowhere near what it is today.

Point J indicates the CO_2 concentration obtained with the 600,000,000 year historical average minimum temperature of 22C, or 1,319 ppm.

Point K shows the resulting temperature of 23.16C using the maximum concentration for CO_2 permitted by OSHA for human indoor habitation of 5,000 ppm.

Point L shows the temperature of 24.79C obtained when the CO_2 concentration 600,000,000 years ago was 7,000 ppm. Even, if we burned every pound of known fossil fuel reserves overnight, we would only be able to get to 45% of where CO_2 was 600 million years ago.

Point M shows the temperature at 1 million ppm: about 323K.

All this does not take into consideration the effects of water vapor, or any other Greenhouse Gases (GHG). Consider, again, that water vapor versus CO_2 (Chapter 2) based on its spectral absorption capability and quantity in the atmosphere has 1352 times more energy absorption capability. Might water vapor be a more significant factor than CO_2?

It does not take into consideration the much more significant warming contribution of solar irradiance or the cooling contribution of changing of cloud cover, or change in the heat reflectivity/absorption of the land and oceans (albedo). It does not take into consideration diffusivity of gases, solar magnetohydrodynamics, and numerous other transfer and transport mechanisms of gases and heat. Nor, does it take into consideration that as atmospheric CO_2 increases or decreases, the absorption rates of the oceans and land are not static and will also change.

Also, the calculated values developed may be considered debatable and subject to revision. However, what is of most importance is the *order* in which these values occur.

Historical Temperature and CO_2

One of the most revealing histories of temperature and CO_2 are presented in the composite graph below. These were shown separately in their respective previous chapters. Both records cover a period of about 600,000,000 years. The CO_2 record is from R. A. Berner's, GEOCARB III Project. The temperature record is from C. R. Scotese's PALEOTEMP Project.

Let's analyze the graphs below to see if there is any correlation between temperature and CO_2. For about 67% of the time during this period temperatures were about 22C, and for about 7% of the time temperatures hovered near the low of about 12C. Right now, we are at 14.5C about 25% off the bottom of the historical temperature range. At one time or another while at 22C, the CO_2 concentration has been at 6800, 2800, 1600 and 800 ppm.

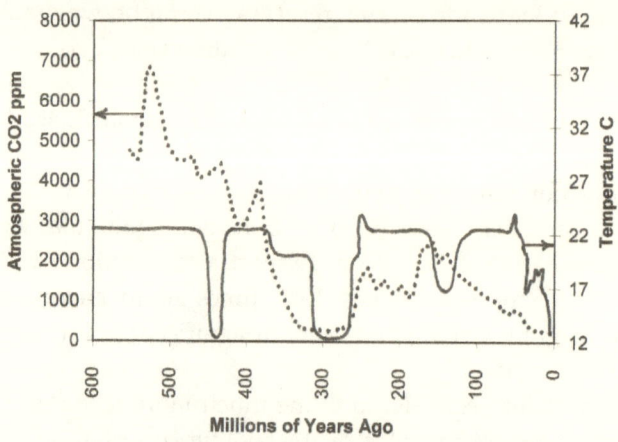

Fig 6c: Temperature Deviation and CO2 Concentration
For The Last 600,000,000 Years
(based on R. A. Berner and C. R. Scotese)

Correspondingly, when temperatures have been at 12C, the CO2 concentration has been at 300 ppm on two occasions and 4300 ppm. During a brief period when temperatures were at about 15C, CO2 was at 2200 and 400 ppm.

Question: What conclusions can be drawn from these relationships? Answer: Historical temperature records have nothing to do with CO2 concentration over the last 600,000,000 years. If you know any AGW worshipers, ask them to explain why there is absolutely no correlation between CO2 concentration and temperature for the last 600,000,000 years. Don't hang by your thumbs waiting for an answer.

<u>Fossil Fuel Use versus Temperature</u>

Another argument made by the global warming alarmists is that fossil fuel burning is increasing CO2 concentration which is driving up temperatures. I do not know what scientific basis they are using to make their case. Perhaps it is the sight of polar bears atop ice floes, a disappearing glacier here or there, a three week heat wave over a particular location, or whatever. But, anecdotes are not proof. They are simply anecdotes. Nothing more, nothing less.

To refute this global warming argument that fossil fuel burning is causing global warming is so easy it is almost shameful. Take a look at the graph above. The dotted line is the fuel consumption in millions of tons between 1860 and 2000. The solid line is the temperature anomaly, or the

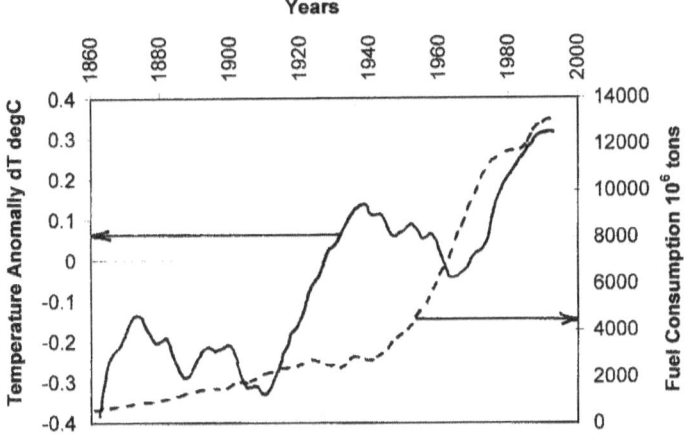

Fig 6d: Fuel Use and Temperature, 1860 to 2000
(Klyashtorn and Lyubushim, 2003 E&E, **14**, 6, Figure 1)

increase or decrease in temperature over the same period. Note how the fuel use is almost always rising. No surprise here. Yet, over this same period there are three increases and two decreases in temperature anomaly. In other words, regardless of how much fossil fuel is burned, the temperature anomaly is obviously being driven by something else.

Temperature and Cosmic Ray Flux

Shaviv & Veizer have also studied the influence of Cosmic Ray Flux (CRF) on Temperature. Their research goes back a little further than 1955. It covers more the half a billion years, which is about 10,000,000 times more in duration than the Sloan & Wolfendale study of one solar cycle.

The graph below plots CRF and Temperature. The approximately 135,000,000 year cycle in temperature reflects the passage of the solar system as it transits the spiral arms of the galaxy. It becomes immediately obvious to even the most casual observer that there is a distinct relationship between CRF and temperature. When the CRF (dotted line) goes up, the temperature (solid line) goes down. Further, the average change in temperature is about 3C or 5.4F. I don't call this change 'a very small part in global warming'. Do you? Compared to the approximately 0.5C rise in global temperatures since the start of the Industrial Revolution, I call this CRF temperature-induced increase as substantial.

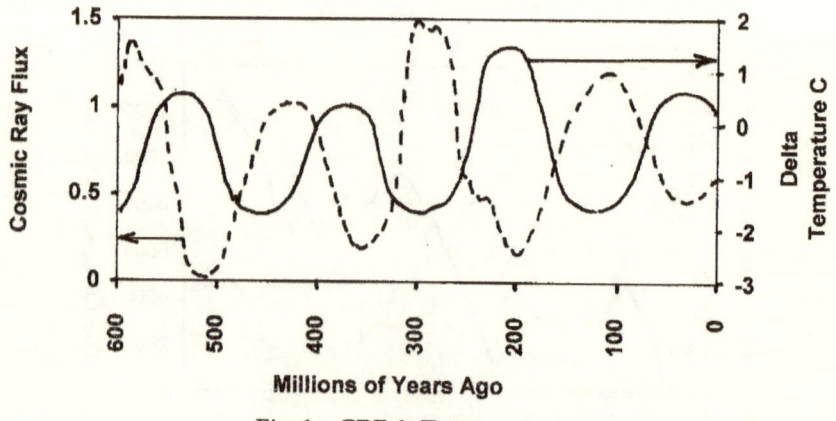

Fig 6e: CRF & Temperature
(Based on Shaviv & Veizer)

Sunspot Cycle and Temperature

Sun spots do not cause anything, but are visible evidence of other phenomena which do alter our climate. When the frequency of this cycle changes, it changes our climate.

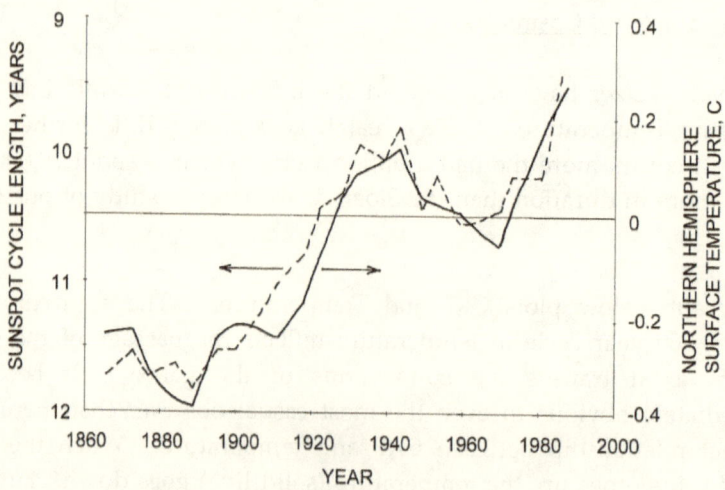

Fig 6f: Sunspot Cycle Length & N. Hemisphere Surface Temperature

As you can see from the above figure, as the length of the sunspot cycle increases northern hemisphere surface temperatures decrease. Notice the peak in temperature that occurred in the decade of the 1930's. This decade is still home for 25 of the all-time record high temperatures recorded in the 50 United States.

Is there any doubt as to where the warming of the Earth has come from in the last 200 years? It is not CO2, which is still near its historic 600,000,000 year low. It is clearly registered from the increase in sun spot numbers.

Solar Irrdiance and Temperature

The IPCC is also intransigent about the effects of cloud cover on the climate. In this regard, it doesn't take a PhD to correctly conclude that increasing cloud cover reflects more sunlight back into space and result in less heating of the atmosphere.

The IPCC is also reticent to acknowledge recent findings from the well respected European Organization for Nuclear Research (CERN). CERN has conducted tests in a cloud chamber to study the possible link between cosmic ray flux and cloud cover. This research expands upon the research initially conducted by Henrik Svensmark and Nigel Calder

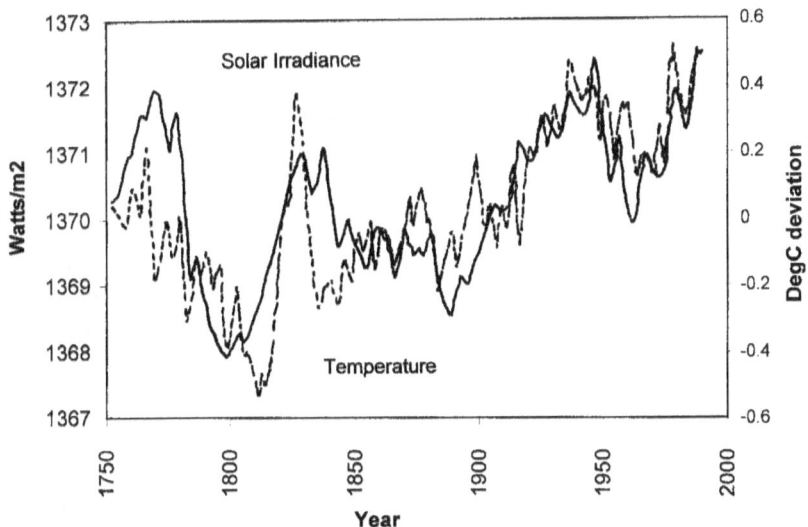

Fig 6g: Solar Irradiance & Atmospheric Temperature Deviation
<SolarActivity_A_DominantFactorInClimateDynamics.pdf>

documented in their book (Svensmark H and Calder N, The Chilling Stars, Icon Books, Canada, 2007).   Unfortunately, CERN's director General Rolf-Dieter Heuer has prohibited CERN scientists from drawing conclusions about this experiment <http://www.theregister.co.uk /2011/07/18/cern_cosmic_ray_gag/>. There is a close correlation between solar irradiance, or how brightly the Sun shines, and the Earth's atmospheric temperature.   This conclusion should be self-evident to almost everyone of any age. AGW alarmists , however, would have you believe that changes in the Sun's output is insignificant and that CO_2 alone is responsible for everything that happens here on Earth.

Earthquakes and Sunspot Cycle

Dr. Charles Davison of England studied records of earthquakes in the northern hemisphere between 1305 and 1899. He discovered during this period the frequency of earthquakes averaged 10.96 years. What Dr. Davison also found was a close correlation between earthquakes and sunspots.

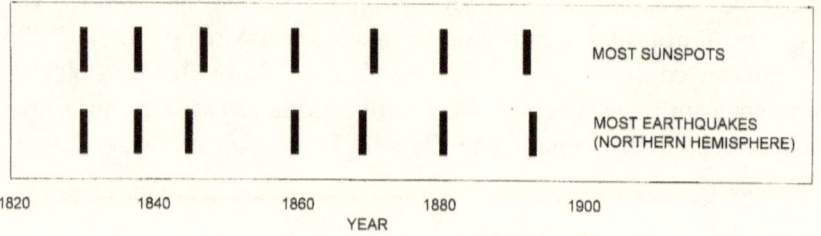

Fig 6i: Earthquakes and Sunspots
(Cycles, Page 141)

Volcanoes and Solar Flares

It has been observed that there is a correlation between solar flares and increased volcanic activity on Earth. There also appears to be an inverse relationship between volcanic activity and the position of Saturn and Jupiter.

As these two Jovian giants reach perihelion (their closest distance to the Sun) as they flew past the Earth, they slow the movement of magma in the Earth's depths. This, in turn, leads to a decrease in the number of eruptions.   <http://tallbloke.wordpress.com/2011/06/18/sun-wakes-volcanoes-up-jupiter-makes-them-sleep/>.

Between 1500 to 1980 it was discovered that two weak cycles of about 11 and 8 years were detected. Both cycles appear to correlate with well-known cycles of solar activity. The phasing is such that the frequency of volcanic eruptions increases or decreases slightly around the times of solar minimum or maximum. (Stothers, R.B., 1989: Volcanic eruptions and solar activity. *J. Geophys. Res.*, **94**, 17371-17381, doi:10.1029/JB094iB12p17371)

Temperature Increases Happen AFTER CO2 Increases

In Al Gore's book and movie, <u>An Inconvenient Truth</u>, Al acknowledges the ice-age long records of correspondence between CO2 and temperature. However, what Al is ignorant of, or more likely, prefers to conveniently forget is the relationship between the CO2 levels and temperatures recorded in ice cores that span the current IWP.

As can be seen in the following graph, CO2 increases occur about 600 to 1,000 years AFTER the temperature increases. How can this be? We are always being told that it is global warming that is driving the CO2 levels to ever higher concentrations. But the data recorded in ice cores reveal precisely the opposite is happening.

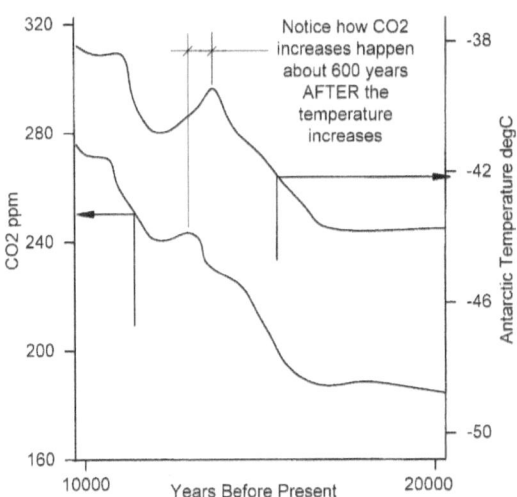

Fig 6h: CO2 Increases AFTER Temperature Increases
(Fischer, H. et al, Ice core records of atmospheric CO2 around the last three glacial terminations, Science 283, 1712–1714 (1999)).

The explanation for this apparent disparity is simple. Just like seeing increasing vapor above a pot of water reaching the boiling point, it is the increasing water temperatures that drive CO2 out of solution from the oceans into the atmosphere.

Conclusion

In 100 years with the present rate of CO2 discharging into the atmosphere from the burning of fossil fuels, I calculate temperature increase at 1.764C. This increase would be about one half of the 3.6C temperature peak experienced during the Minoan Warming Period 4,300 years ago. So, what's the problem?

"It's only when the tide goes out that you learn who's been swimming naked."

– Warren Buffett

Chapter 7: Ocean Cycles

One of the most powerful influences on our climate is the cycles of cooling and warming directly attributable to the oceans. The oceans contain 329,000,000 cubic miles of water, or about 1.45×10^{18} or 1,450,000,000,000,000,000 tons. The total weight of the atmosphere is about 5.5×10^{12} or 5,500,000,000,000 tons. Thus, there is about 260,000 times more mass in the oceans than there is in the atmosphere. Also, the specific heat of water is about 4 times that of air. Therefore, the heat retention capability of the oceans is about 1,000,000 times more than that of the air. So, from a thermodynamic standpoint, the oceans are the 800 pound gorilla in the room, not the atmosphere.

Oceans and Seas

The major bodies of water around the world consist of five oceans: Atlantic, Arctic, Indian, Pacific, and Southern (Antarctic). There are smaller bodies of water such as seas, bays, gulfs, straits and a channel or two, here and there. From medieval times the seven seas included the Adriatic, Arabian, Black, Caspian, Mediterranean, Persian and Red Sea. There are about 40 other seas. Some are simply forgotten about and considered part of an ocean, like the Sargasso Sea, which is the Bermuda triangle portion of the Atlantic. Some seas are land locked like the Aral, Caspian, Dead, Salton and Black Seas.

Ocean Gyres

An ocean gyre is vortex, a large system of circulation currents brought about by the Coriolis effect of the Earth. So, just like the atmosphere, the oceans are affected by the Coriolis effect. The five main gyres are the North Atlantic, South Atlantic, Northern Pacific, Southern Pacific and Indian. The two gyres north of the equator rotate clockwise and the three south of the equator rotate counterclockwise.

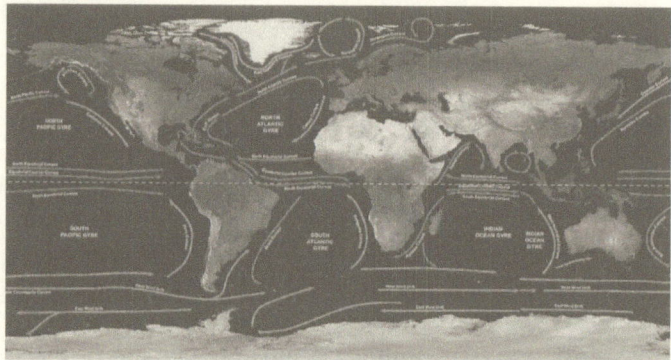

Fig 7a: Ocean Gyres
(Courtesy of NOAA)

Besides the five ocean gyres, there are ocean currents. These currents are classified as either surface currents or deep water currents.

Ocean Surface Currents

Ocean surface currents are caused by solar energy, gravity and wind. These currents redistribute the vast amount of solar energy absorbed by the Earth's Oceans from equatorial regions towards the poles. Gravity pulls colder ocean waters below warmer waters. Wind, if persistent like trade winds and westerlies, leave their presence as relatively consistent patterns of surface water flows. The combination of these influences result in the Ekman Effect.

The Ekman Effect was discovered by Vagn Walfrid Ekman in 1902. He found that the frictional force of the winds combined with the Coriolis effect of the Earth's rotation impart a transport of water 90 degrees from

84

the direction of the wind.   The direction of the transport, like atmospheric cyclones, is clockwise in the northern hemisphere and counterclockwise in the southern hemisphere.

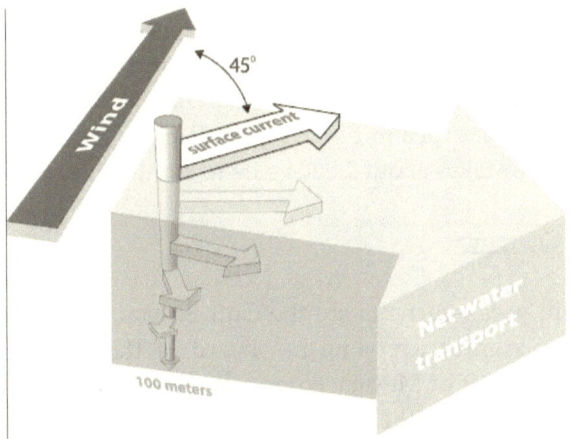

Fig 7b: Ekman Effect
(ocean service.noaa.gov)

Ocean Deep Water Currents

Ocean deep water currents, on the other hand, are controlled by the Great Ocean Conveyor.   This name was coined by Dr. Wallace Smith Broecker in his 1991 paper. (Broecker, W. S., The great ocean conveyor, *Oceanography* 4(2):79–89, https://doi.org/10.5670/oceanog.1991.07).   This diagram first appeared as an illustration in an article about the Younger Dryas event that was published in the November 1987 issue of *Natural History.*

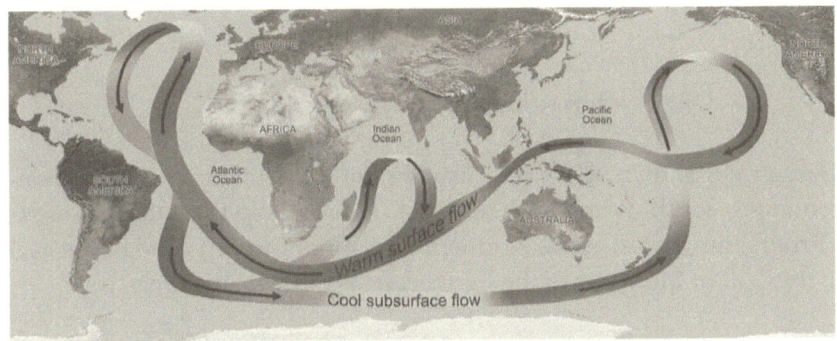

Fig 7c: Great Ocean Conveyor

The Great Ocean Convey is also called the Thermohaline Circulation (THC). As the name implies, the THC is a transport mechanism powered by the differences in temperature of warmer and cooler water and density differences between fresh melt water and salt water. The temperature differences occur in two areas, the North Atlantic Deep Water (NADW) flowing between Greenland and Great Britain into the Northern Atlantic Ocean, and the Antarctic Bottom Water (AABW) flowing from the Ross Sea to the Southern Pacific Ocean. It is estimated that this conveyor takes about 2,000 years to complete a round trip.

Main Ocean Currents

The four main ocean currents are the Gulf Stream, California, Labrador and Agulhas. Almost everyone has heard of the Gulf Stream which originates in the Gulf of Mexico, comes up the East Coast of the US, and heads over to Europe via the North Atlantic Drift.

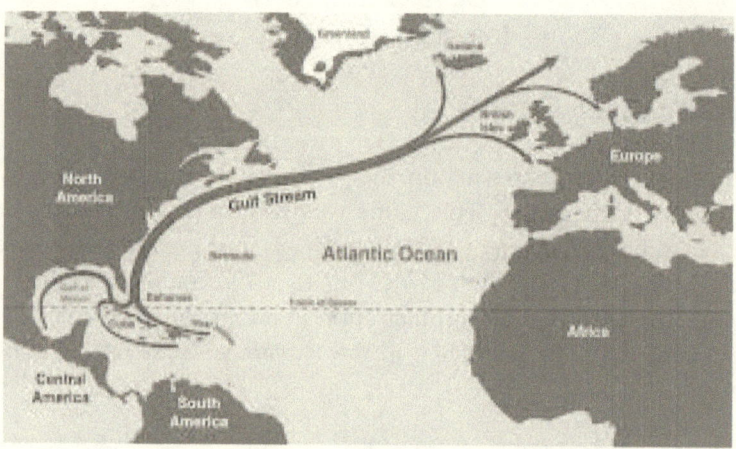

Fig 7d: Gulf Stream
<https://scijinks.gov/gulf-stream/>

Though not shown above and usually not mentioned, the Gulf Stream continues south skimming the west African coast as the Carnaries Current , turns west as the Northern Equatorial Current , and heads back to the Gulf of Mexico thru the Caribbean to complete the cycle.

The California Current is a Pacific Ocean current that moves southward along the western coast of North America, beginning off southern British Columbia and ending off the southern Baja California Peninsula.

The Labrador Current begins as the southward East Greenland Current turns northward as the West Greenland Current, makes a U-turn in Baffin Bay and heads south along the Labrador Coast. Eventually, the Labrador Current meets up with the Gulf Stream off the coast of Newfoundland. It is at this junction that the Titanic met its fate with an iceberg delivered by the Labrador Current.

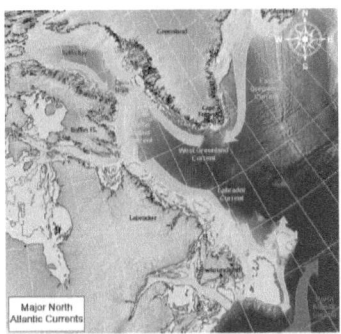

Fig 7e: Labrador Current
<https://navcen.uscg.gov>

The Agulhas Current rides down the east coast of Africa and turns sharply east as the Agulahs Return Current of the coast of Capetown. The South Atlantic Current heading east also turns north here to become the Benguela Current. Below this is the Antarctic Circumpoler Current traveling east. The confluence of these currents is the Subtropical Convergence and results in numerous eddys and rogue waves. This area is a notorious burial ground for many ships. If you see videos of 100 foot rogue waves pounding ships or ships sinking, there is a good chance they were taken in this turbulent, violent area.

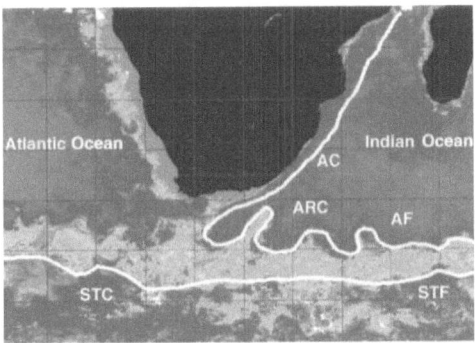

Fig 7f: Agulas Current
https://sealevel.jpl.nasa.gov/images/ostm/science/invest-provost-fig3.html

Arctic Ocean

To listen to the alarmists one is lead to believe there is this body of water around the North Pole with an ice cap whose very existence is held by a thread, continually threatened by mankind. A simple look at the following figure shows the complexity of just the flows of water taking place into and out of the Arctic Ocean.

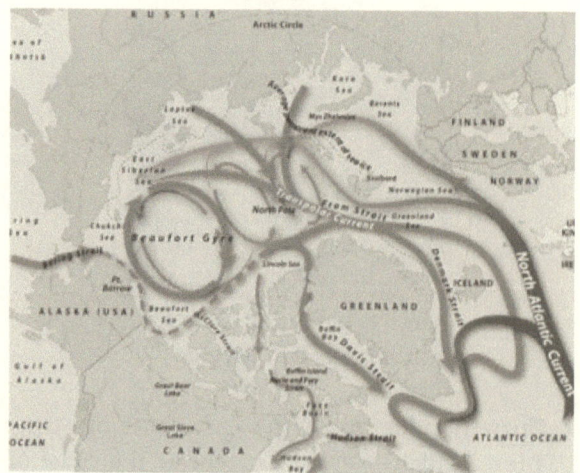

Fig 7g: Arctic Ocean Currents
(Courtesy of NOAA)

The seas surrounding the Arctic Ocean include the Chuckchi, E. Siberian, Laptev, Kara, Barents, and Beaufort. Direct saltwater inflows come from the Bering Strait and North Atlantic Current. Direct freshwater inflows come from the Amor, Anadyr, Kolyma, Indigirka, Lena, Yenisey, Ob, Pechora, Sev Dvina, Lagen, Koksoak, LaGrande, Eastmain, Nelson, Churchill, Thelon, Back, Copper-mine, Mackensie, Colville, Yukon, Copper, Stikine, Skena, and Frazer Rivers.

The primary influence on the Arctic Ocean is from the North Atlantic Current. This current supplies the bulk of water, heat and saltwater into the Arctic Ocean, enough to melt the ice sheet many times over. However, this inflow is not constant.

Around 1990 temperature measurements recorded by submerged buoys anchored hundreds of meters below the surface revealed about a 1C warming pulse. A second warm water pulse was recorded starting around 2004. (Polyakov et al, Fate of Early 2000s Arctic Warm Water

Pulse, AMS, Vol 9, No. 5, May 2011). It was during these warm water pulses that the record-breaking ice minimum of 2007 was observed which became the *proof* the warmistas needed that Climate Change (CC) was reaching epoch proportions. After 2008, the warming pulses ceased and a cooling period of about 1C was being recorded.   As in the past, these hot transients dissipated and things returned to normal. Questions should be posed as to where do these pulses come from?   Are they cyclical? Can they be predicted?

Pacific Decadal Oscillation (PDO)

One of the most important contributors to the weather is the PDO. Since the Pacific Ocean occupies about 1/3 of the Earth's surface, this should come as no surprise. The PDO directly controls the kind of winters and summers that will be experienced by the United States. Indirectly, it's effects extend to other circulations around the world.

Amazingly, this potent driving force of weather and climate was not even named until 1996. It was then that a fisheries scientist, Steven Hare, at the University of Washington noticed regular decadal shifts in the relative concentrations of certain fish species. It was during the cool phases along America's west coast that fishermen obtained high anchovy yields.   When the oscillation reversed and became warm, the anchovy yields dwindled but the sardine yields increased.

In 1998, NASA proclaimed the beginning of the cool phase of the PDO. This cooling is expected to last another 30 years. Prior to 1998, the PDO was in its warm phase, which accounted for much of the world's alleged global warming. Al Gore and other worshippers at the church of AGW will shun this revelation, since it does not support the tenets of their religion.

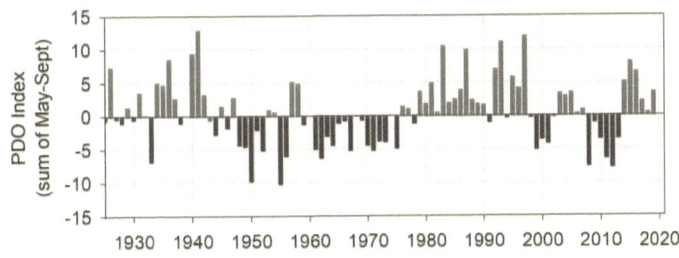

Fig 7h: Pacific Decadal Oscillation – 1925-2019
<https://www.nwfsc.noaa.gov/research/divisions....pdo.cfm>

Using tree ring proxies, the PDO has been reconstructed going all the way back to the year 993. The PDO has a 50-70 year cycle.

Atlantic Multidecadal Oscillation (AMO)

The AMO was only recently identified in 1994. It appears to correlate with temperature and rainfall variations in North America and Europe, and is related to droughts in the US Midwest and South. The AMO has a cycle of about 25 to 40 years.

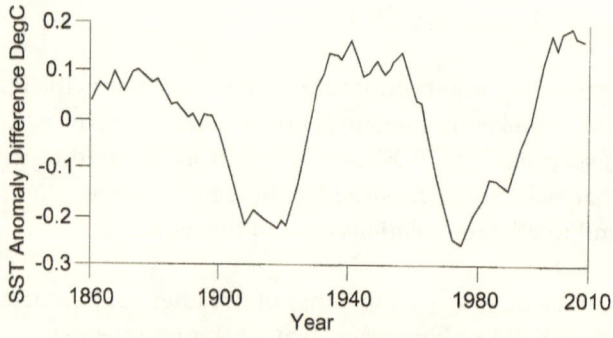

Fig 7i: AMO – 1860-2010

< www.usgs.gov/media/images/atlantic-oceans-warm-vs-cool-cycles>

Indian Ocean Dipole (IOD)

The IOD, also known as the Indian Niño, is an oscillation of sea surface temperatures in which the western Indian Ocean becomes alternately warmer and then colder than the eastern part of the ocean. This alternation occurs about every 30 years. Within this period there are four each intervening warm peaks and cold peaks with no apparent pattern.

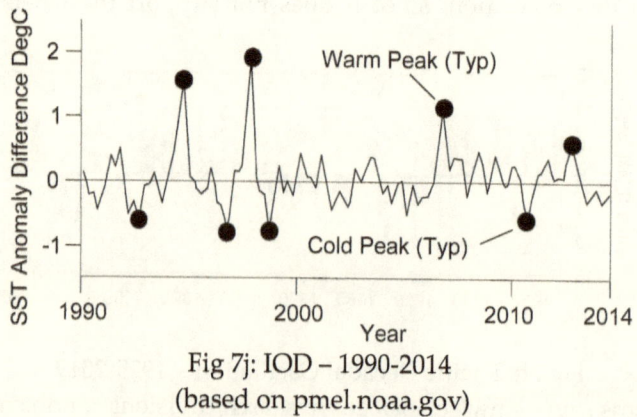

Fig 7j: IOD – 1990-2014

(based on pmel.noaa.gov)

The IOD is one of the key drivers of Australia's climate and can have a significant impact on farming and is responsible for many droughts, which frequently lead into their infamous bush fires. The IOD has three phases: neutral, positive and negative. Events usually start around May or June, peak between August and October and then rapidly decay when the monsoon arrives in the southern hemisphere around the end of spring.

The IOD was only named in 1999. However, its monsoon season is well known and was the subject of a movie <u>The Rains Came</u> with Tyrone Power in 1939, which was remade in 1955 as <u>The Rains of Ranchipur</u>. The IOD greatly affects the rainfall of East Africa.

<u>Summary</u>

- Ocean surface currents are created by solar, gravity and wind influences.

- Ocean deep water currents are created by the Great Ocean Conveyor or Thermohaline Circulation.

- The THC controls the ocean oscillations.

- The ocean oscillations control the atmospheric circulations.

- The three most influential ocean oscillations were discovered only within the last couple of decades.

"All great truths begin as blasphemies."

- George Bernard Shaw

Chapter 8: Atmospheric Cycles

Everyone is aware of the regular arrival of cold weather in winter, hot weather in summer and temperate weather every spring and fall. No surprise here; these are seasonal cycles. Then there are regular cycles of dry and rainy weather which happen every three days, or so, the result of passing high and low pressure areas. Once-in-a-while there can be a period of extended rain and drought. These are usually the result of a blocking high pressure ridge, halting the parade of highs and lows. These are the most commonly known atmospheric weather cycles. But there are many other cycles, whose effect leave their subtle and not so insignificant mark on our weather.

The 15-30 Day mountain induced torque cycle

A 15 to 30 day atmosphere circulation pattern has been identified which is the result of the mid-latitude jet stream interacting with large scale topography (mountains). (Lott et al, Mountain torques and atmospheric oscillations, GRL, Vol. 0, No. 0, 2001. Additionally, these atmospheric oscillations have been shown to affect the Arctic (Ocean) Oscillation (AO) and Pacific North Amercian (PNA) ocean pattern.

30-60 Day Madden-Julian Oscillation (MJO)

The Madden-Julian cycle was discovered by Roland Madden and Paul Julian in 1971 & 1972. It is the largest element of the intra-seasonal (30-90 day) variability in the tropical atmosphere. It is a pattern which travels eastward between about 4 and 8 meters/second through the atmosphere <en.wikipedia.org/wiki/Madden-Julian_oscillation>. The MJO is also known as the 30-60 day oscillation, 30-60 day wave, and intra-seasonal oscillation and takes about 40-50 days to circle the globe.

The MJO has been found to affect large differences in Atlantic hurricane frequency and intensity. Its effect can be projected about two weeks out and compliment already available longer term projections. (Klotzbach, Philip J., On the Madden-Julian Oscillation-Atlantic Hurricane Relationship, Journal of Climate, Vol. 23, page 282, 2010)

433 Day Chandler Period Atmospheric Oscillation

The Chandler Wobble period of about 433 days has been found to cause an atmospheric tide at the 700 hPa level. (Benuzzi, Eugene Joseph, Masters Thesis: Chandler period atmospheric oscillations at the 700 Hecto Pascal level over the Northern Hemisphere, 20 Dec 1978). So, even though Chandler's Wobble, the wobble of the Earth's rotation axis, is a meager few meters, it's effects on the atmospheric pressure are large enough to be measured.

26 Month Tropospheric Heat Engine Cycle

A distinct 26-month oscillation in the troposphere has been named the Tropospheric Heat Engine Effect. The signature of this biennial effect has been found to influence surface temperature, rainfall, tree rings, lake levels, ozone levels and other conditions which depend on the troposphere. The source for this oscillation is the apparent change in diameter of the Sun. (Newell, Reginald E., Nature **204**, 278-279, 17 October 1964). However, of the other few papers written about the 26-month heat engine cycle, none acknowledge a changing solar diameter as an influence. (Smith, C.; Messina, D., In: Sun and planetary system; Proceedings of the Sixth European Regional Meeting in Astronomy,

Dubrovnik, Yugoslavia, October 19-23, 1981. (A82-47740 24-89) Dordrecht, D. Reidel Publishing Co., 1982, p. 39-43. NASA-supported research.)

The Younger Dryas

The Younger Dryas has been identified as one of the most abrupt climate changes in Northern Hemisphere (Brauer et al, Nature Letters, **Vol 1**, 520-523, August 2008). The data gathered shows a period of storminess due to a stronger and more zonal jet stream. This abrupt climate change occurred between the autumn and spring of year 12,679 BP and was a wind-driven atmospheric event. This was the last major cold shot before the recent Little Ice Age of 300 years ago, interrupted by the Minoan, Roman and Medieval Warming periods.

Fig 8a: Younger Dryas Event Evidence
<http://www.physorg.com/news106410997.html/cometmayhave.jpg>
Credit: Allen West, University of California, Santa Barbara

The photo above shows a distinctive black layer which marks a line of extinction that occurred about 12,900 years ago.

Equatorial Electro Jet Path

The equatorial electro jet (EEJ) is the name given to a band of intense electric current which flows eastward around the equator at an altitude of between 90 and 130 km. The band is about 600 km wide and follows the path shown in the figure below. It was discovered in 1920's from magnetometer measurements taken in Huancayo Peru. The magnetic strength of this field measures about 165 nanoTeslas (nT).

The source and movement of the EEJ are not that well understood. It is theorized that this jet is the result of changes in the solar wind, the magnetosphere, wind shear, and solar-lunar interactions. The three-letter designations above are abbreviations for observation stations used to measure the electrojet as it meanders around the Earth.

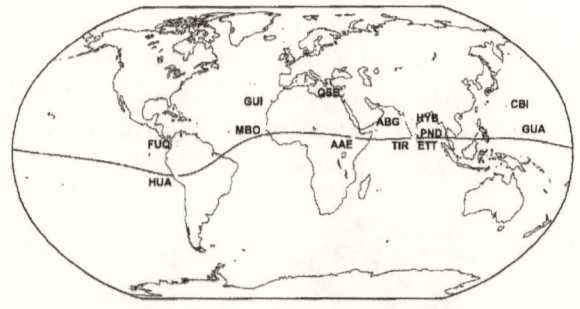

Fig 8b: Equatorial Electrojet Path Around Earth
(Reddy, C. A., PINSA 64, No. 3, May 1998, 353-364)

Teleconnections

The circulation of the atmospheric varies greatly. This occurs on many time scales from a few days in summer followed by a stormy cold front, to a few weeks like a persistent Bermuda high off the US East Coast, to a few months, to several years, or even centuries. Teleconnections refer to a recurring and persistent, large-scale pattern of pressure and circulation anomalies that spans vast geographical areas.

For the ocean oscillation graphs the y-axis is simply the Sea Surface Temperature (SST) of the water. For teleconnections the y-axis is an index based on Rotated Principal Component Analysis. Say what? If you want the details and be bored to death, check out the Climate Prediction Center website.

NOAA

The National Oceanic and Atmospheric Administration (NOAA) was originally formed in 1807. It is charged with monitoring and analyzing the Earth's oceans and atmospheric systems. It has six major line offices including The National Ocean Service and The National Weather Service. NOAA's teleconnections web page lists five teleconnections including the Antarctic Oscillation (AAO), Arctic Oscillation (AO), El Nino Southern Oscillation (ENSO), Pacific Decadel Oscillation (PDO), and Pacific North Atlantic (PNA) Oscillation.

NOAA's products web page lists climate and weather links including the AAO, AO, NAO and PNA, and El Nino La Nina and MJO.

NOAA's Data webpage lists ten oscillations contained within the North Atlantic, Eurasian and North Pacific/North Atlantic areas. Are your eyes bleary yet? Rather than fill pages endlessly, let's look at the most influential of all these atmospheric oscillations.

Arctic Oscillation (AO)

The AO involves a seesaw pattern in atmospheric pressure between the North Pole and middle northern latitudes. Its negative phase brings higher than normal air pressure over the Arctic region and lower than

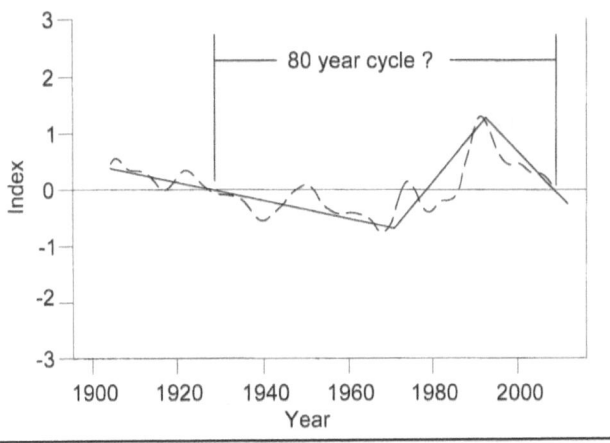

Fig 8c: Arctic Oscillation Index 1900-2010
<www.cpc.ncep.noaa.gov/products/precip/CWlink/daily_ao_index/JFM_
season_ao_index.shtml>

normal pressure over the central Atlantic Ocean. These pressure differences lead to weaker westerly winds north and south of the equator. North of the equator, the weak westerlies allow cold Arctic air to reach farther south. During the cool phase, much of the U.S., as well as Northern Europe and Asia, experience cold and stormy winters. In the extreme, this condition is referred to as a visit from the Polar Vortex. The AO's positive (warm) phase brings about the opposite conditions, with much of the U.S. and Northern Europe experiencing mild winter weather.

At first glance the AO appears to have a high frequency cycle only lasting a few of years as per the dashed line in the graph above. But looked at over a span of decades it appears to possibly take on a low frequency cyclical trend with a period of about 80 years.

Antarctic Oscillation (AAO)

The AAO is a ring of variability that encircles the South Pole and extends as far north as New Zealand. It is characterized by a seesaw pattern in atmospheric pressure between the Antarctic region and the middle southern latitudes. The AAO positive phase brings relatively light winds

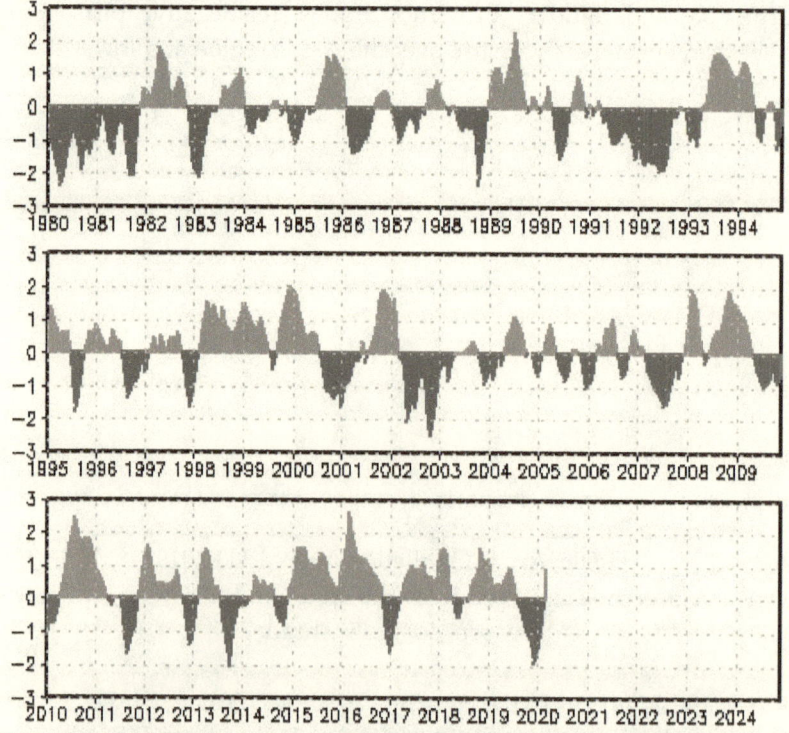

Fig 8d: Antarctic Oscillation Index - 1980 to 2020
<www.cpc.ncep.noaa.gov/products.../aao/month_aao_index.shtml>

and more settled weather to middle latitudes, together with enhanced westerly winds over the southern oceans. The AAO negative phase brings stronger westerlies over middle latitudes, with more unsettled weather, while windiness and storm activity ease over the southern oceans.

The AAO is a high frequency oscillation that lasts about only two years or less as per the three graphs above.

North Atlantic Oscillation NAO

The NAO is a periodic change in atmospheric pressure between Iceland and Portugal that affects the strength of prevailing westerlies over the North Atlantic Ocean. This produces the strongest influence on weather patterns over the Northeast U.S. of any of the oscillations. These winds, in turn, affect the strength and direction of surface currents in the North Atlantic. The NAO positive mode involves high atmospheric pressure developing over the Azores, and an intense low over Iceland. This results in stronger ocean winds and milder winters in the eastern U.S. The NAO negative mode has weaker ocean winds and more severe U.S. winters.

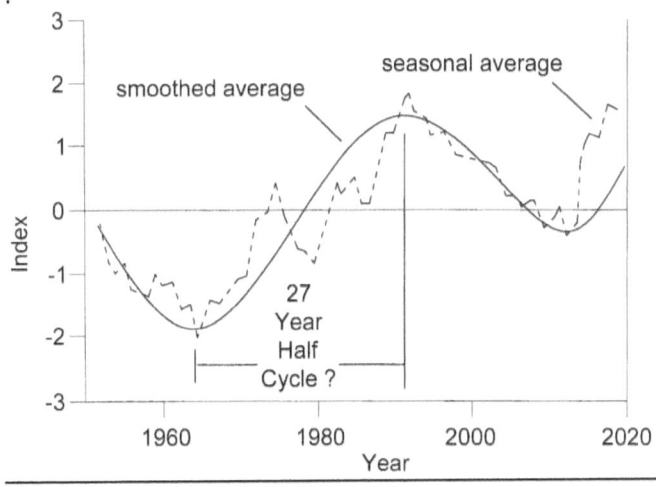

Fig 8e: North American Oscillation - 1950 to 2020
<www.cpc.ncep.noaa.gov/products/....../JFM_season_nao_index.shtml

The NAO is a low frequency cycle. The smoothed average half cycle of about 27 years may suggest a 54 year cycle.

Pacific North American Oscillation (PNA)

The PNA Pattern relates the atmospheric circulation pattern over the North Pacific Ocean with the one over the North American continent through the height of the sea surface in the northern Pacific. The changes in sea-surface heights causes strong fluctuations in air pressure and temperature here and to the East. The PNA positive mode is associated with changes in the strength and location of the East Asian jet stream, above-average temperatures over western Canada and the western U.S., and below-average temperatures and drought conditions across the south-central and southeastern U.S. The PNA negative mode involves a westward retraction of the East Asian jet stream, and a strong split-flow configuration over the central North Pacific. The western U.S. may experience relatively cold and wet conditions, while the eastern U.S. remains warm and dry during these negative modes.

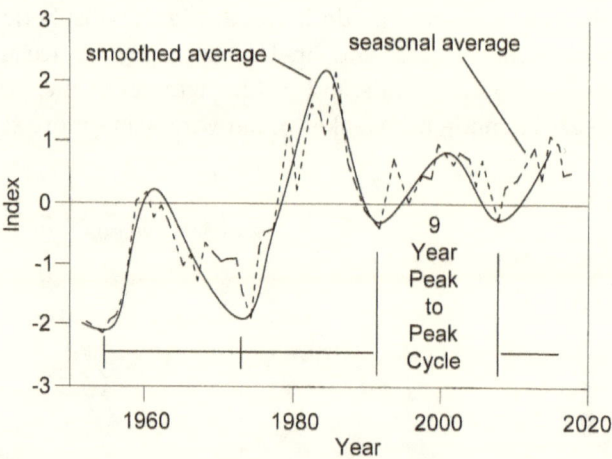

Fig 8f: Pacific North American Oscillation – 1950 to 2020
<www.cpc.ncep.noaa.gov/products/.....JFM_season_pna_index.shtml>

The PNA seems to have a low frequency oscillation of about 9 years peak-to-peak. This duration may suggest the influence of the lunar nodal cycle of 18.6 years.

El Nino and La Nina

Without question, the most familiar of these ocean influences are El Nino and La Nina. Even with the daily bombardment of manmade global warming nonsense put forth by the MSM, they still manage to get things right sometimes and tell us what really is causing those dry spells, wet spells, snowy winters, etc. It isn't global warming, it's El Nino, or La Nina.

Before 1960, no one ever heard of El Nino or La Nina, yet they have a long history. Chemical analysis of corals indicate that warmer sea surface temperatures and rainfall caused by El Ninos date back more than 100,000 years. Five hundred years ago, fisherman off the coast of Peru were aware that warmer waters would reduce their catch of Anchovies. In 1891, Dr. Luis Carranza published an article associating El Ninos with rain patterns. In 1923, Sir Gilbert Walker discovered a southern oscillation of air pressure between the Pacific and Indian Oceans.

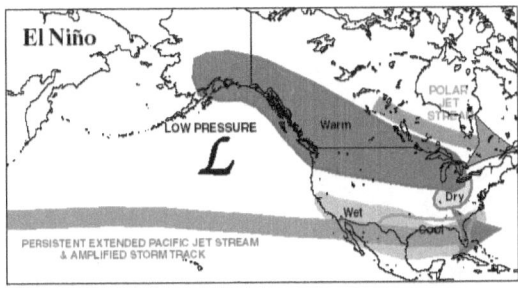

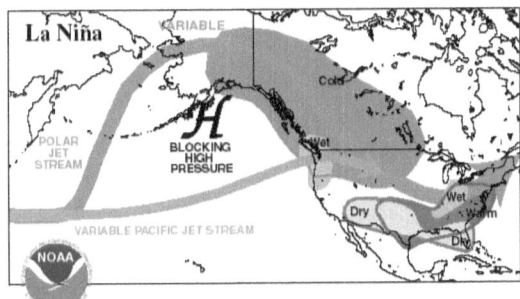

Fig 8g: El Nino and La Nina
(Courtesy of NOAA)

El Nino Southern Oscillation ENSO

In 1969, Professor Jacob Bjerknes at UCLA formulated the first detailed description of how the ENSO functions. But, it wasn't until 1982-1983, where the strongest El Nino up until that time was recorded, that attempts to study this phenomenon really began. Then in 1997-1998, an even more powerful El Nino made its appearance causing havoc in the southeast US. Not until recently did El Nino become part of our daily lexicon of terms we use frequently to describe changes in climate today, a natural phenomenon even acknowledged by the Church of Global warming.

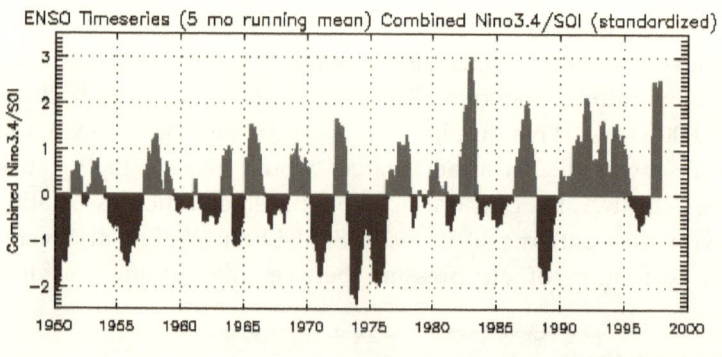

Fig 8h: ENSO Timeseries
(NOAA-CIRES CDC)

For the graph above, depending on who is doing the counting, there are 27 cycles of the ENSO timescales over a period of 50 years. Depending on who is doing the observing these cycles can vary from 1 year to almost 6 years in duration. So, while there is definitely a cyclical time series, the cycle durations seem to be rather chaotic.

Conclusion

Isn't it amazing that - as potent as these atmospheric oscillations are, which have so much of an influence on our weather and climate - they were only discovered within the last few decades?

"Pretending to know everything, closes the door to finding out what's really there."

- Neil DeGrasse Tyson

Chapter 9: Temperature Cycles

Since 1998

Since 1998, there has been a statistical global cooling of the Earth's atmospheric temperature as measured by satellite. This cooling corresponds with the cooling of the Pacific Decadal Oscillation and a decrease in Total Solar Irradiance. So, while the CO_2 keeps increasing, two of the true drivers of global cooling keep decreasing along with attendant satellite temperatures.

Much to the dismay of the worshipers in the church of anthropogenic global warming, temperatures have not been increasing, recently. This difficulty was acknowledged by Phil Jones of Climate Research Unit in one of the famous Climategate emails in which he stated "The scientific community would come down on me in no uncertain terms, if I said the world had not cooled from 1998. OK, it has but it is only 7 years of data and it isn't statistically significant". Climatologist Kevin Trenberth in an email dated 10-12-09 said: "The fact is that we can't account for the lack of warming at the moment and it is a travesty that we can't". To these two pontiffs in the church of AGW, I submit that the problem might be that your global warming theory might be WRONG.

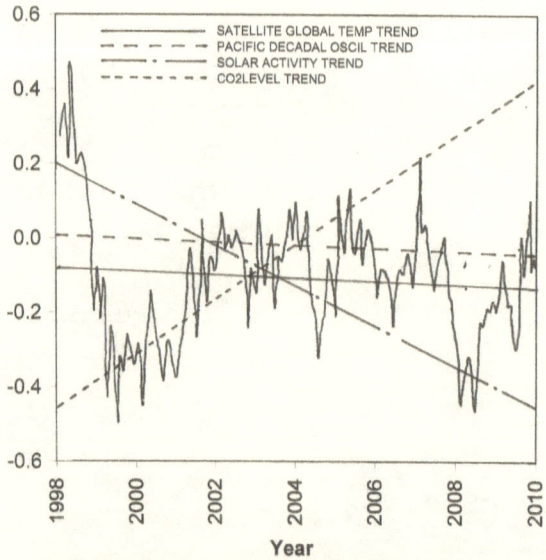

Fig 9a: Temperature ForThe last 17 years
<http://hockeyschtick.blogspot.com/>

The graph above is devastating to the church of AGW. It clearly shows, that while CO2 marches steadily upward, temperatures march steadily downward in sync with Total Solar Irradiance, Pacific Decadal Oscillation and the Satellite Global Temperature Trend. I know, I know, this is only 10 or so years of data. To which I agree at first glance should amount to nothing. But, what happens when the window of observation is increased?

The Last 140 Years

One of the most telling temperature records of the last 140 years is the all-time record high set in each of the 50 States in the USA. With all the recent talk of 'this year' being the hottest year on record, it's time for a wake-up call.

Since 1880, each of the 50 States in the USA have set an all-time record high temperature as shown in the bar chart below. Notice that there are no hottest records for the 1900s, 1940s, 2000s or 2010s. Also, note that 25 of the 50 records were set in the decade of the 1930s. This should be no surprise to anyone as this was the decade of the Dust Bowl as depicted in the 1939 novel <u>The Grapes of Wrath</u>.

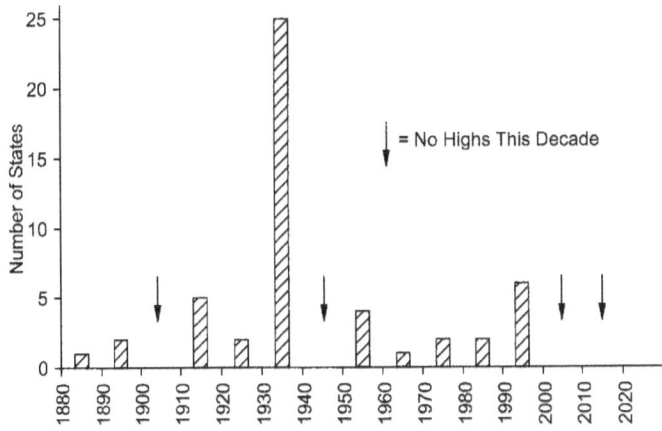

Fig 9b: Hottest Decades for the last 140 Years
<www.infoplease/math-science/weather/hottest...>

If you go to Wikipedia, there is a slight difference in these data. One State that stands out is Colorado. At Wiki the record is 115F set at the John Martin Reservoir July 20, 2019. This location is indicated with a "note [3]", which refers to an article that appeared in CPR News. CPR is Colorado Public Radio located in Centennial CO. <https://www.cpr.org/2019/10/04/its-official-colorados-wild-weather-year-sets-new-high-temp-and-hailstone-records/>. The article goes into details about how the news of this new historic temperature reading first appeared in the "Twitterverse" [Getting nervous, yet?]. The article mentions the 110F limit of the recording drum in the thermograph, and attempts by some individuals to get corroborating temperature readings. You can go to the article to get the rest of the details. For me, Wikipedia is an OK go-to source for non-controversial, inane topics like art, entertainment, sports, geography and the like. But when it comes to politics and global warming, I would suggest you go elsewhere. In the meantime, I'll stick with the Infoplease numbers.

The Last 250-years

During the last 250 years, there have been 6 distinct cycles of alternating cooling and warming. These cycles march to the tune of Total Solar Irradiance. The odds of this happening by chance is 2^6, or 1 out of 64. Thus, we can be 98.43% certain that this correlation is no coincidence.

Add to this curve the statistical cooling that has occurred since 1998, and a seventh cycle can be added bringing the odds up to 1 out of 128, or 99.22% (Scafetta, Nicola, Climate Change and Its Causes, ArXiv:1003. 1554v1 [physics.go-ph] 8 Mar 2010)

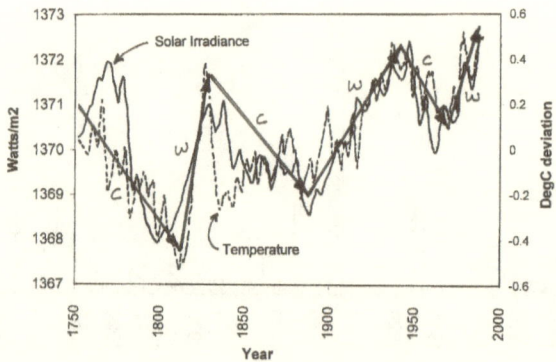

Fig 9c: Temperature Deviation For The Last 250 Years
(based on Fig 11 of Scafetta, 8 Mar 2010)

The light solid curve is the driving function of solar irradiance; the dotted curve is the atmospheric temperature deviation. It is clearly obvious that within the last 240 years alone there have been six distinct warming and cooling trends, noted by the very dark arrowed lines, up and down marching in lock step with solar irradiance. For the IPCC to conclude that the effect of solar irradiance on our climate is insignificant requires total ignorance of the facts, recognition of an agenda, or both. So, who is in denial here?

The last 1,200 Years

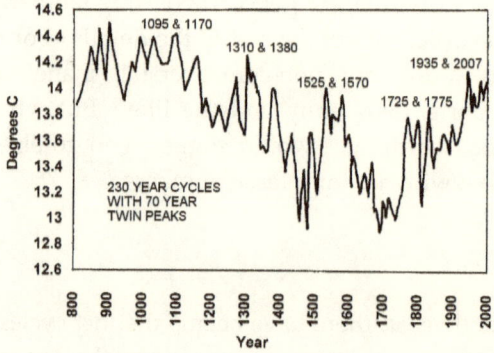

Fig 9d: Temperature Deviation For The Last 1,200 Years
(Global Weather Oscillations - Natural Climate Pulse 2007)

106

During the last 1,200 years there have been five 230-year temperature cycles. Each one of these 230-year cycles has a 70-year sub-cycle with double peaks. Can this be a coincidence in that double peaks also occur in the 11 year sunspot cycle?

The Last 415,000 years of Vostok Ice Core Data

So, the AGW crowd is alarmed at the recent 0.5C rise within the last 200 years. How would they react, if they found out that hundreds of thousands of years ago it was eight and ten times warmer than the increase in temperature since the start of the Industrial Revolution?

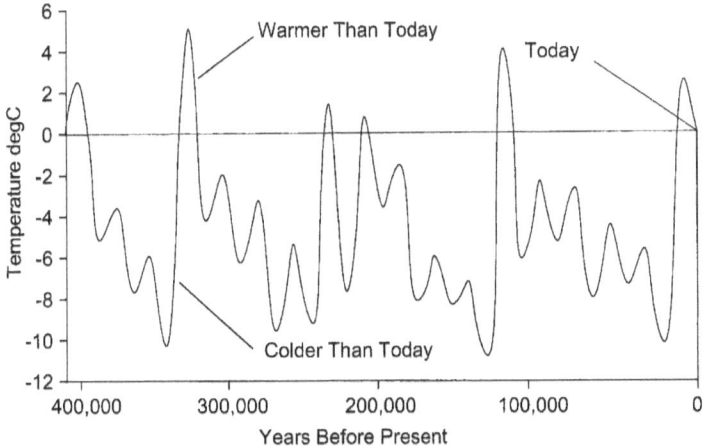

Fig 9e: Vostok Ice Core Temperatures for The Last 415,000 Years
(JGR 1998, Vol 103, no D8, pages 8963-8977)

Take a gander at the graph above. Notice at about 120,000 years ago temperatures were 4C higher than today, and that about 330,000 years ago temperatures were 5C higher than today.

Incidentally, isn't it curious how there are two, three or four ever-decreasing peaks into temperatures colder than today followed by the one increasing peak where it was warmer than today? Doesn't this curve suggest some cyclical - possibly planetary - influence at work here?

Summary

- Since 1998, there has been no warming at all according to satellite temperature measurements.

- In the last 140 years, 25 of the 50 all-time hottest State temperatures were recorded in the 1930s, and none in the 2010s or 2020s.
- Within the last 250 years there have been three cooling and three warming cycles.

- About 120,000 years ago temperatures were 4C higher than today. About 330,000 years ago temperatures were 5C higher than today.

- Within the last 550,000,000 years temperatures have varied in a range between 12C and 22C. Today we are at 14.5C, only 25% off the bottom of the range.

- For the last 150 years there has been no correlation between fossil fuel use and CO2 generation.

"The most exciting phrase to hear in science is not "eureka" but "that's funny""

- Isaac Asimov

Chapter 10: Time Cycles

The 4.33 Year US NE Cities Precipitation Cycle

The precipitation records of New York City, Philadelphia and Baltimore between 1820 and 1960 share a common thread. The cycle of precipitation is 4.33 years. Chen and Shi at the Department of Atmospheric Sciences, Nanjing China after analyzing the CRU global precipitation dataset have discovered a 2-7, 13 and 20 year cycle of worldwide precipitation.

The 7.54 Year NYC Barometric Pressure Cycle

Since 1873 the barometric pressure of New York City has cycled at 7.54 years. A barometric pressure study of weather stations around the world produced similar results of 7.54 years.

The 22.75 year Great Lakes Water Level Cycle

The water levels of the Great lakes are controlled by precipitation, run off and evaporation. In 1964 the Great Lakes reached their lowest level in history. This may not sound important, but lower water levels can be devastating on lake shipping, fishing and hydro power generation.

Colonel Harry A. Musham of Chicago predicted in 1941 that Lake Michigan would reach a maximum in 1952. In 1943 he predicted it would reach a minimum in 1963. He was right on both counts. Musham's work indicates that the Great Lakes have a water level cycle of about 22.75 years.

20 Year NAO Salt Water Concentration Cycle

A 20 year cycle has been discovered in the salt water concentration of the North Atlantic Oscillation (NAO) . (Perez, F.F, et al, Scientia Marina, Vol 64, No. 3, 2000, Climatological coupling of the thermohaline decadel changes in Central Water of the Eastern North Atlantic)

20 Year Northern China Drought Cycle

A 20 year oscillation in drought has been isolated after studying the 163 year history of tree ring widths of pine and spruce trees in Northern China. (Liang ErYuan et al, SpringerLink Journal Article, 2000).

40 Year Cycles

From NASA, the Met Office, NOAA and the Japanese Meteorological Agency there is a distinct 40 year cycle of temperature.

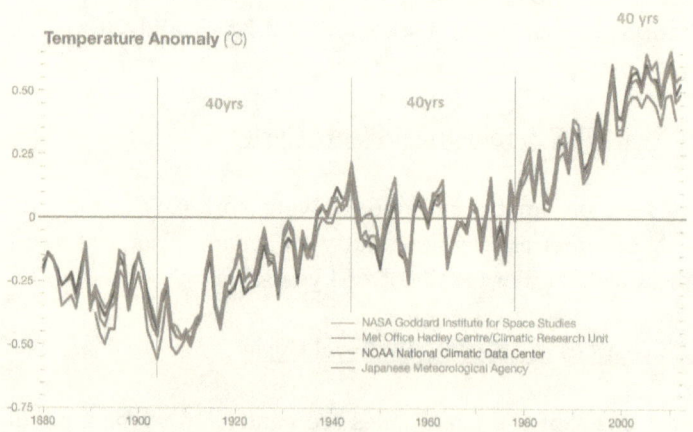

Fig 10a: Temperature Anomaly 1880 to 2010
<go.nasa.gov/climate365>

60 Year Cycle

The 60 cycle is registered in numerous scientific, economic, cultural aspects, including temperature, sea level, Chinese Calendar, Kondratieff Waves, stock market, sea levels, climate, solar irradiance, etc.

The 60 cycle in temperature has an amplitude of 0.3C, implying that 60% of the 0.5C increase claimed by the IPCC is the result of natural causes. (Nicola Scafetta, arXiv:1003.1554v1, 2010)

A 60 year cycle of temperature, solar irradiance and galactic cosmic rays has been discovered between temperature, and solar irradiance and the galactic cosmic rays over the last 150 years. (Komitov, Boris, The Sun-Climate Relationship: III, The Solar Eruptions, North-South Sun Spot Area Assymetry and Climate, Bulgarian Astronomical Journal, vol 13, 2010)

It has been discovered that there is a 60 year cycle in Global Mean Sea Level (GMSL) (Chambers, D. et al, Is there a 60-year oscillation in global mean sea level?, GRL, Vol 39, No 18, 2012)

The Chinese Calendar Stem Branch Cycle is also named the sexagenary cycle or the stem-branch cycle, and finds its fist recorded use over 2,000 years ago. The Chinese 60-year calendar cycle is based on the combinations of a cycle of ten heavenly stems and twelve earthly branches. Each year is named by a pair of one stem and one branch.

The 100 Year Wet/Dry Cycle

Raymond H. Wheeler while at the University of Kansas spent much of

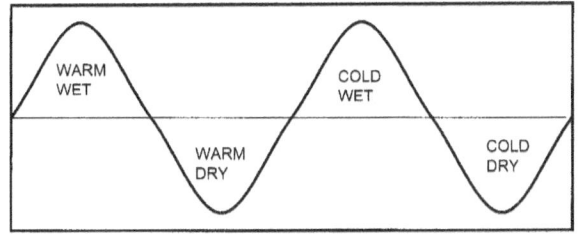

Hundred Year Cycle
Fig 10b: Wheeler's Phases in Climate
(based on Cycles, Page 137)

his life compiling a history of events. Professor Wheeler discovered many cycles in the course of his career, but his most important discovery was the hundred year cycle. This cycle has an average length of about 100 years, but can be as short as 70 years and as long as 120 years. This hundred year cycle marches to a sine wave, and repeats itself every cycle.

Dr. Wheeler also developed the Drought Clock, which predicted droughts using his 100 year cycles together with a 170 year and 510 year cycle. The conjunction of these three cycles occurred in 450, 975, 1460 and 2000.

Harris-Mann Climatology has taken this cycle to the next level of detail. The first lobe is a warm wet phase characterized as the 'best' cycle with its prosperity and brief peace during 1893 thru 1917 and 1995 thru 2019. The second lobe is a warm dry phase and is the 'worst' period known for war and depression during 1917 thru 1943 and 2019 thru 2045). The third lobe is a 'good' cold wet phase being artistic and religious during 1943 thru 1968. The fourth lobe is a 'bad' cold dry phase considered migratory and selfish during 1968 thru 1995. This cycle is predicted to reach a very unfavorable completion cycle of 6,120 years due by 2038.

The 400 Year Wet/Dry Cycle

This 400 year cycle is attributed to solar forcing as indicated by changing levels of cosmogenic isotopes of samples taken in sites from Minnesota to Alberta. (Zichen Yu and Emi Ito, The 400-year wet-dry climate cycle in Interior North America and its Solar Connection, 2002 PACLIM Conference Proceedings). Similar results were obtained of samples taken in Northwest China. (Wu, J., Yu, Z., Zeng, H. *et al.* Possible solar forcing of 400-year wet–dry climate cycles in northwestern China. *Climatic Change* 96, 473–482 (2009). https://doi.org/10.1007/s10584-009-9604-4)

The 500 Year Cycle

A 500-year monsoon activity cycle spanning the last 8000 years has been detected in Northeast Asia. The attendant warm-humid/cold-dry phases of monsoon cycles track closely with levels of human activity and prosperity. These cycles appear to be linked to the El Niño-Southern Oscillation. (Deke Xu et al, Synchronous 500-year oscillations of monsoon climate and human activity in Northeast Asia, <u>Nature Communications</u> volume 10, Article number: 4105 (2019)

<u>The 725 Year Drought Cycle</u>

This 725 year drought cycle oscillation has been recorded from 400 to 4900 BP and seems to correlate with Indian Ocean Monsoons. (Russell et al, A 725 yr cycle in the climate of central Africa during the late Holocene, Geology, August 2003, v.31, no.8, p 677-680).

<u>The 1470 Year Dansgaard-Oeschger Warm Event Cycle</u>

To quote from the paper: "Dansgaard-Oeschger (D-O) cycles are the most dramatic, frequent, and wide-reaching abrupt climate changes in the geologic record." <agupubs.onlinelibrary.wiley.com/doi/full/10.1002/palo.20042>

About 11,500 years ago, or the start of the present IWP, the averaged annual temperatures on the Greenland ice sheet increased by around 8 °C over 40 years. That's 14.4F, or 0.36F per year! Talk about global warming.

And all this temperature increase happened without any fossil fuel burning or human influence.

"There is something out there . . . that affects every living thing on earth, and it does so with rhythms that have taken man through cycles of war and peace, prosperity and depression, optimism and despair, discovery and isolation, morality and degradation, creativity and ignorance, famine and plenty "

– Edward R. Dewey

Chapter 11: Abundance Cycles

Edward R. Dewey was the Chief Economic Analyst at the Department of Commerce in 1931. At that time he was asked by President Hoover to discover "why a prosperous and growing nation had been reduced to a frightened mass of humanity selling apples on street corners" after the Crash of 1929. In Dewey's search for this answer he came across numerous cycles, some of them economic and many others having nothing to do with the economy. He found rhythms in nature, production of goods, patterns of war, cycles in the universe. He found cycles in church membership, guilt, cigarette production, steel production, airplane traffic, marriage rates, death rates, crime, precipitation, barometric pressure. He found cycles in the prices of corn, wheat, cotton, pig iron, and oats. He found rhythms in stock prices. He discovered relationships between sunspots and manufacturing. (Dewey, Edward R., Cycles – The Mysterious Forces That Trigger Events, Hawthorn Books, 1971.

Ultimately, he saw the planetary relationships, and discreet yearly cycles of abundance, such as, the 5.91, 8, 9.2, 9.6 and 18.2 yearly cycles. The most important observation Dewey made was that all these cycles turn at the same time. That is, with dozens of different abundance cycles of, say, 9.6 years they all increase and decrease together.

One of the most prolific cycles is that of 9.6 or 9.7 years. This period covers 37 listings within the subjects of mammalogy, Ichthyology, Ornithology, Entomology, Dendrochronology, Agronolmy, Climatology, Hydrology, Medicine, Sociology and Economics.

Dewey asserted that seemingly unrelated time series often had similar cycles periods present and that when they did the phase of these cycles was mostly very similar (cycle synchrony). He also said that there were many cycles with periods that were related by powers or products of 2 and 3. This is illustrated in the table below.

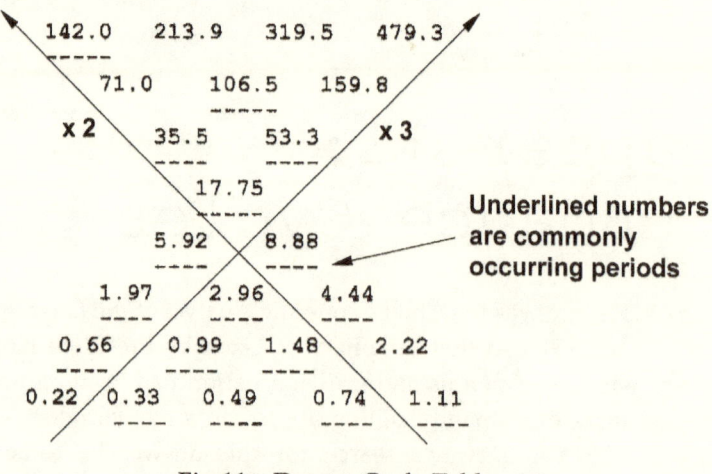

Fig 11a: Dewey Cycle Table
(based on Cycles, Page 137)

To construct this table starting from the period 17.75 years, multiply by three as you proceed along diagonals from lower left to upper right, and multiply by two as you proceed along diagonals from lower right to upper left. Dewey reported that the underlined numbers are commonly occurring periods (in years).

Cycles Research Institute

If you go to the Cycles Research Institute web page for cycles/cycles index there is a table of cycles ranging from 1 hour and 4 minutes for the Old Faithful geyser to 510 years for civil wars. In this table you will find over a thousand cycles including rhythms in nature, wars and battles, commodities, production, prices, sales and abundance.

Rhythms in nature

Ozone records of the Greenwich Observatory in London and Montsouris Observatory in Paris, between 1876 and 1906, show a definitive 9.6 year cycle of ozone concentration.

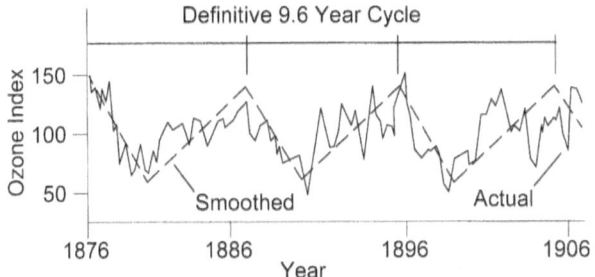

Fig 11b: Ozone Measurements, London & Paris, 1876 to 1906
(Based on Cycles, 2012, Page 56)

Going back to Fig 2e which covered ozone concentration between 1958 and 1987, you wonder if the Ozone Trend Panel is aware of the very cyclical data above between 1876 and 1906. Want to venture a guess? My wager is that the OTP is aware, but cannot acknowledge it. Otherwise, their ozone hole fear mongering agenda goes right out the window.

Variations in the electrical charge of the atmosphere and the growth of Sequoias in California also have a similar cycle. The abundance of lynx, tent caterpillars, chinch bugs, coyotes, grouse, hawks, martens, minks, muskrats, owls, salmon, skunks, snowshoe rabbits, and wolves varies by 9.6 years. This 9.6 cycle is prevalent in nature and is roughly half that of the lunar nodal cycle of 18.6 years. Lunar cycles appear to be the driver of abundance in insects, as well.

Wars and Battles

Wars appear in cycles of 11.25, 33 and 50 years. Civil wars at intervals of 170 and 510 years. There are 43 listings for battles. There are over a hundred listing for wars.

Planting Cycles

It has been found by gardeners that it is important to plant particular crops at particular times. Annual plants produced above ground are best planted during the first or second quarter of the Moon, preferably close to the full or new moon. Root crops should be planted just after the full moon. <http://www.gardeningbythemoon.com/calc.html>

Fishing and Hunting Cycles

John Alden Knight developed a theory in the 1930's on which are the best times to go hunting or fishing to obtain maximum results. The major periods occur when the combined gravitational forces of the sun and moon are the strongest. These occur when the Sun is over head or under foot. The minor periods occur when the sun and moon are passing the horizon. These periods also correlate with the high and low tides. The best hunting and fishing times also are aligned with phases of the moon, with peaks occurring around the Full and New Moons. <http://i-solunar.com/i-Solunar-Theory.php>. Yes, and they have an "app" for that, too.

*"At the top of the **cycle** you write [policies for] everybody, no matter how bad, and at the bottom you cancel everybody, no matter how good. It's a manic-depressive **cycle.**"*

- Robert Hunter

Chapter 12: The Sun

The atmosphere, hydrosphere, and cryosphere are to the greatest degree affected by our Sun. These spheres are represented by the gaseous, liquid and solid states of water. These states are the result of evaporation, condensation, freezing and melting. Just enough energy is furnished by our Sun to keep us at the narrow Goldilocks band of temperatures allowing these changes of state all at one atmosphere pressure.

<u>Our Sun</u>

Our sun is a massive chemical factory which continuously converts hydrogen into helium. It is a fission reaction which, fortunately for us, has been virtually constant for millions of years. There are subtle changes in output which is noticeable at the surface. Total Solar

Irradiance (TSI), for example, varies about every 60 years. This variation only amounts to 1 or 2 watts per square foot as measured at the Earth's surface from its average of 1370 w/sf.

Due to its extreme brightness, special filtering techniques are used to observe the Sun. We all learned from early childhood never to look directly at the Sun. To do so would cause blindness. To get around this hindrance, indirect observation methods were developed. One of three methods consisted of making a pinhole in a box and allowing the Sun's image to be cast onto the rear of the box to be viewed, indirectly. Later, when telescopes allowed magnified viewing of the sun, apertures were placed in front of the telescopes to drastically cut down the amount of light being collected to allow direct viewing. Today, much of the viewing of the Sun is conducted by satellites. Below is a photograph of the Sun taken by NASA's Solar and Heliospheric Observatory (SOHO) satellite. I can only guess what the "f stop" number on the sensors were. Maybe f1000?

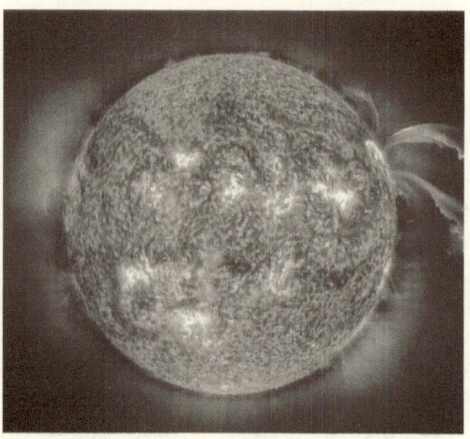

Fig 12a: The Sun
<http://sohowww.nascom.nasa.gov/gallery/images/
large/eit002_prev.jpg>

All this tugging away at the Sun has an effect of just more than constantly shifting the center of the Sun around. The other effect of all this planetary motion is the change in angular momentum. This is the Law of Conservation of Angular Momentum. The result of this causes the surface velocity of the sun at the equator to change differently from that near its poles. This twisting action felt by the Sun is just like the reaction felt by a wet rag being wrung out. In the case of the rag it is water that comes out. In the case of the Sun something else comes out. This 'something else' is solar wind, flares, eruptions and coronal mass ejections.

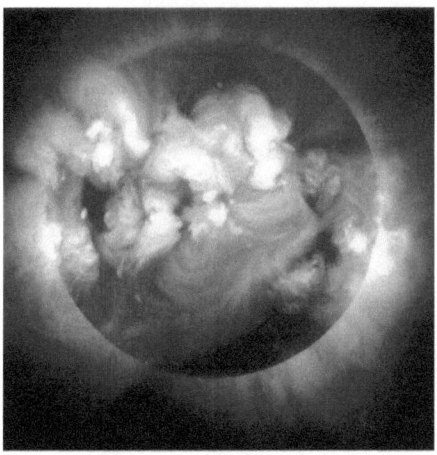

Fig 12b: Solar Eruptions
<www.nasa.gov/images/content/462158main_solar1.jpg>

The sun is not just a ball of hot gas. It has very distinct layers. The core where the fusion reaction of hydrogen to helium takes place is about 100,000-km in diameter and operates at 27 million degrees F. Next, is the radiative zone which is about 300,000-km thick. This is followed by the convective zone of about 200,000-km thickness. Then there is the photosphere, which is the surface we see. It is covered mostly by granules with some sun spots. Granules are cells about 1,000-km wide that carry energy from below to the surface. The temperature of the photosphere is about 10,000F.

Above the surface is the atmosphere which itself has layers, including the chromosphere, transition layer, and finally the corona. The chromosphere rises in temperature to 36,000F. Above the chromosphere is the transition layer where temperatures increase up to 500,000F. The outermost atmospheric layer is the corona ,which gets up to 2,000,000F. It is in the corona we get to see the prominences, flares, filaments and streamers during eclipses. It is from the corona where the solar wind begins. The "p" and "g" notations have to do with sound wave propagation within the Sun. And you thought those "ear buds" could cause hearing loss. Can you imagine the decibel level inside the Sun? It would put heavy metal bands to shame.

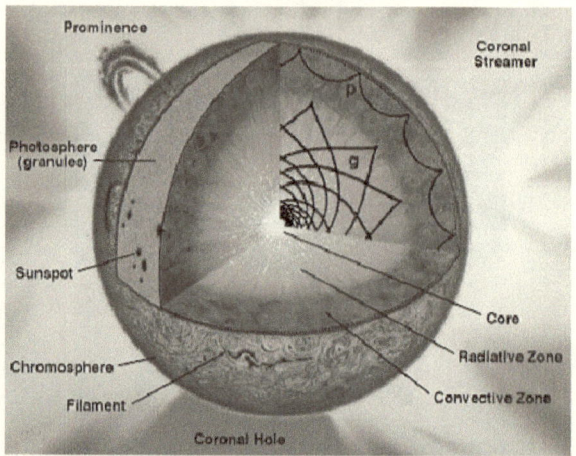

Fig 12c Cross Section of the Sun
<http://sohowww.nascom.nasa.gov/gallery/
Helioseismology/large/mdi004_prev.jpg>

So, the Sun is very hot in the center, comparatively warm at the surface, and moderately hot again above the surface. Logic would expect a gradual transition from hot to cold as the distance from the core increases. Scientists are not sure why this hot-cool-warm transition in temperature takes place.

Personally, I have problem with the temperature distribution estimates by scientists. A core at 27 million degrees and corona at 2 million degrees will not produce a lower temperature in between. This is a heat transfer impossibility; the temperature must be higher not lower. Either the temperature assumptions by science are wrong, or something else, such as, magnetic effects are negating the effects of heat transfer.

Sun Spots

Sun spots are caused by intense magnetic activity, which inhibit convection. This causes areas of reduced temperature which create the dark spots. The magnetic filaments can be seen rising vertically from the perimeter of the dark spots and bending over.

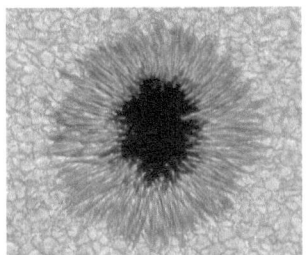

Fig 12d: Sun Spot Closeup
<sohowww.nascom.nasa.gov>

Has it ever occurred to you why Sun spot cycles happen with such regularity every 11 or so years? The Sun does not have a built-in alarm clock or computer. Yet, sun spot cycles occur with uncanny consistency.

Fig 12e: Inactive Sun with almost no Sun Spots
<http://sohowww.nascom.nasa.gov/sunspots/>

The picture above was taken about 11 years ago. In 2020, the sun was a total blank, again.

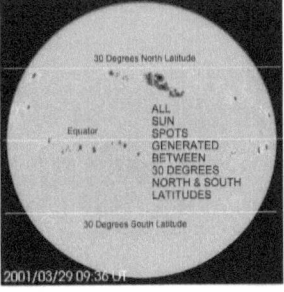

Fig 12f: Active Sun with Sun Spots
<sohowww.nascom.nasa.gov>

Almost all sun spots occur between 30 degrees north and south latitudes.

The figure below depicts how often sun spots have occurred between 1875 and 2010. In each roughly 11 year sun spot cycle, spots are produced in a butterfly pattern. That is, at the beginning of each cycle the spots occur more often near the 30 degree latitudes, and as the cycle progresses to the end more spots are produced near the equator.

DAILY SUNSPOT AREA AVERAGED OVER INDIVIDUAL SOLAR ROTATIONS

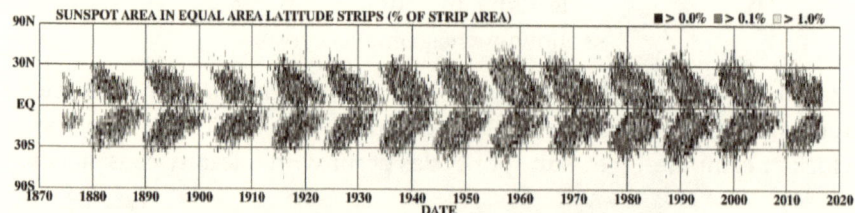

Figure 12g: Sun Spot Butterfly Diagram
< https://solarscience.msfc.nasa.gov/SunspotCycle.shtml >

Notice how the spots before 1900 rarely get outside the 30 degree latitudes, but after 1900 more and more spots are registered outside the 30 degree latitudes as the date approaches the present.

Sun spot activity has been reconstructed over the last 12,000 years. As can be seen below the sunspot number has been higher for the last 200 years than it has been for the last 11,000 years. To be sure, this was not caused by humans or the Industrial Revolution. Remember, that it has

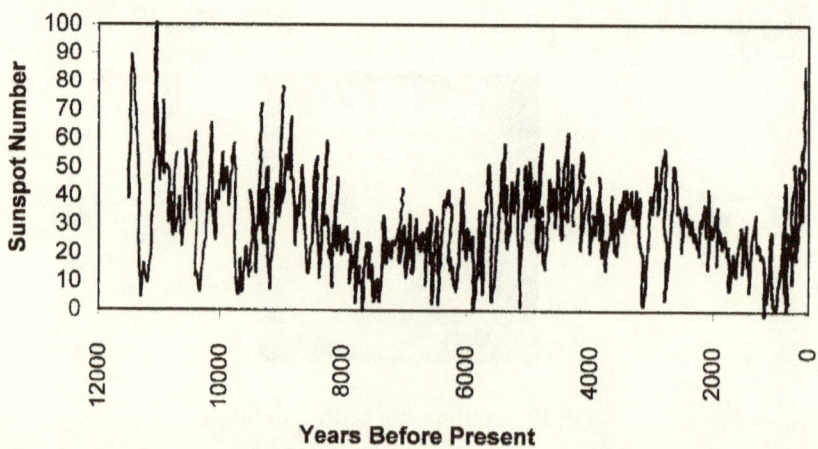

Fig 12h: 12,000 year Sun Spot History
(based on Max Planck Institute For Solar Research)

been only the last 11,000 years where the Earth has enjoyed the balmy temperatures we are enjoying today. Before this, the Earth was in the grasps of the last Ice Age.

Vukcevic

There is evidence of a multi-resonant system within solar periodic activity as discovered by Vukcevic.

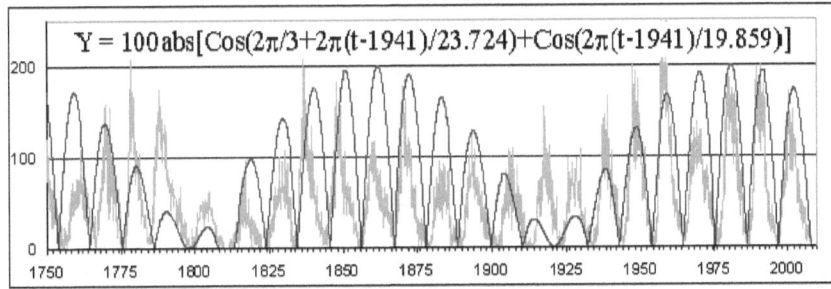

Fig 12i: Multi-resonant sun
(M.A. Vukcevic. Synchronising effect of planetary resonance. 2004. hal-00001357v3)

Jupiter and Saturn account for 92% of the total mass and 86% of the total angular momentum of all the planets. Additionally, Jupiter accounts for 96.76% and Saturn 2.87% of the magnetic field influence of all the planets. So, it should not be any surprise that these two giants have such a great influence on the Sun. It is also of significance that the orbit period of Jupiter and the average sunspot period are so closely aligned.

Increased solar activity results in an increase of harmful radiation, reducing the amount of oceanic plankton by sterilization by irradiation, resulting in reduced uptake of CO_2 from the atmosphere, elevating the 'green-house' effect, increasing the global temperature. Decreased solar activity reverses the process.

The Solar Conveyor

On the Sun similar convection currents are created. Global convection created by the Coriolis force drives the differential rotation of the Sun. Besides the proportionately larger size of the Sun, it is a giant ball of gas, whereas the Earth is a fairly solid body with a skinny coating of atmosphere. Thus, the circulations within the sun reach tens of thousands of miles deep, while on Earth the circulations are only a few dozen miles deep. This plasma flow has been called the Solar Conveyor.

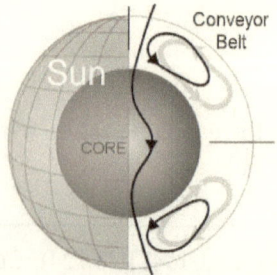

Fig 12j: The Solar Conveyor
<http://science.nasa.gov/media/medialibrary/2007/08/05/
10mar_stormwarning_resources/conveyorbelt.jpg>

From www.nasa.gov/mission_pages/sunearth/science/plasma-flow .html 'The upper belt skims the Sun's surface, sweeping up decaying sunspots and carrying them toward the poles. The structure and strength of this meridional flow is believed to play a key role in determining the strength of the Sun's polar magnetic field, which in turn determines the strength of the sunspot cycles.'

The Solar Dynamo

The solar dynamo is the name given to the changing magnetic fields within the Sun. This dynamo is self-exciting and reverses polarity approximately every 11 years. These magnetic fields are produced by currents generated from hot ionized gases.

The solar dynamo cycle begins with a quiet Sun having only polar or poloidal magnetic field lines coming out of the south pole and bending out into space and back towards the north pole.

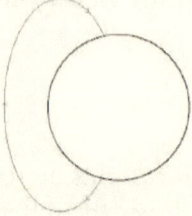

Fig 12k: Sun's Poloidal Magnetic Field
< http://solarscience.msfc.nasa.gov/dynamo.shtml>

As global convection proceeds a differential twisting of convection layer begins which shears the poloidal field into a toroidal field. This is called the Omega effect.

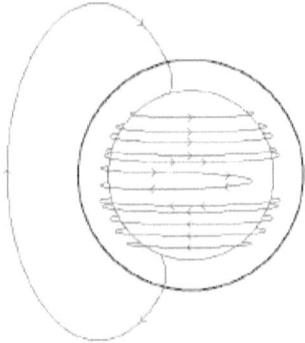

Fig 12l: Sun's Omega Effect and Toroidal Field
< http://solarscience.msfc.nasa.gov/dynamo.shtml>pic

When the toroidal field becomes strong enough, buoyant loops rise to and penetrate the surface giving us sun spots. This is called the Alpha effect, named for the shape of the twisted loops that form in some of the toroidal field.

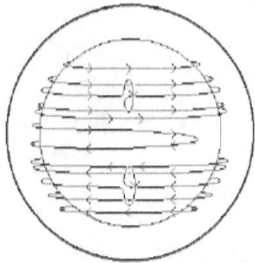

Fig 12m: Sun's Alpha Effect
< http://solarscience.msfc.nasa.gov/dynamo.shtml>

At this point the alpha-effect closes the dynamo loop, and starts regenerating the collapsed poloidal field from the toroidal field. At the same time, the solar conveyor transports the surface magnetic flux poleward, returning some of it on the bottom of the conveyor towards the equator. This returned flux is of opposite sign, where the process begins again with reversed polarity.

To get a better visual appreciation for the polodial and toroidal flows, they should be watched in an animation video such as the one on YouTube <http://www.youtube.com/watch?v=bhr4qiEfSHE>

It should come as no surprise that these magnetic fields manifest themselves the way they do. The Sun rotates about once every 28 days at the equator and about 22 days at the poles. This tremendous twisting action must create circulations. Or is it the circulations that are causing the twisting patterns?

Solar Wind

The solar wind is a plasma of charged particles consisting of protons, electrons, and ionized atoms emitted by the upper atmosphere of the Sun at an average speed of 400 km/sec, or about 1,000,000 mph. The SOHO satellite monitors the solar wind and numerous other properties of the Sun comprising the solar weather. You can monitor all this weather on www.spaceweather.com. SOHO is parked in an L-1 Lagrangian orbit, permanently between the Earth and Sun about 1,500,000 km, or 932,000 miles from Earth. If there is a sudden burst of wind that could be large enough to disrupt satellite communication, SOHO can give about a one hour warning as the wind passes it. Visual observations give about 92 hours notice of possible impacts of the solar wind.

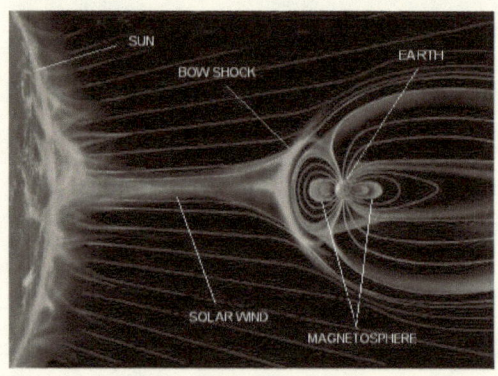

Fig 12n: Solar Wind
<http://www.nationalatlas.gov/articles/geology/IMAGES/solar_wind.jpg>

The solar wind varies in proportion to solar poloidal magnetic field strength, not the number of sun spots. The solar wind speed has also been found to lead cosmic ray flux variation by about two years. Near-Earth variations in solar wind, measured by geomagnetic "aa" index since 1868 have been found closely correlated with global temperature. Further, solar geomagnetic activity leads Earth global temperatures by 4 to 8 years. Thus, solar flare incidences can be used as predictors of temperature changes. <http://landscheidt.auditblogs.com /papers-by-dr-theodor-landscheidt/solar-wind-near-earth:-indicator-of-variations-in-gl obal-temperature>. At the time of the writing of this paper in 2000, the prediction was made that for the next 30 years there would be a cooling of the Earth due to reduced solar activity. So far, for the last 10 years this prediction has come to pass.

Observation

Stop and ponder for a while.

Isn't it amazing that the Sun's fusion reaction marches on with uncanny stability, throughout recorded history, all without instruments, computers, and other things that could otherwise interfere and upset its delicate balance?

Can you imagine mankind trying to replicate this beautiful, powerful machine, without something going terribly wrong?

*"All motion is **cyclic**. It circulates to the limits of its possibilities and then returns to its starting point. "*

- Robert Collier

Chapter 13: The Moon, Earth & Planets

<u>Our Moon</u>

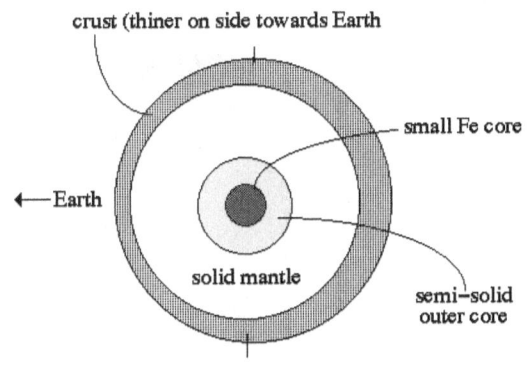

Fig 13a: Lunar interior
<rst.gsfc.nasa.gov/Sect19/Sect19 6a.html>

Isn't it interesting how the interior of the moon is off center with the thinner side of the crust pointed towards Earth. This is due to the rotation of the moon taking the same time to revolve on it's axis as it takes to revolve around the Earth.

Our Earth

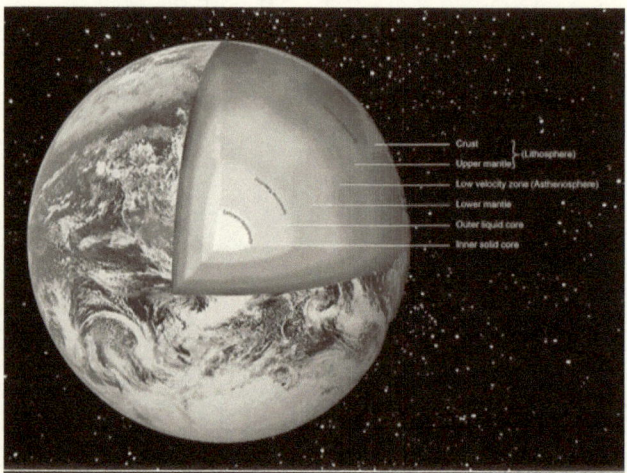

Fig 13b: Earth's Internal Structure
www.ukspaceagency.bis.gov.uk/.../Earth/8385.aspx

The Nine Planets

Lined up, the Sun and nine planets have been characterized as the fingers and thumbs on two hands. The Sun is one thumb with the inner set of

Fig 13c: The Nine Planets of The Solar System
<www2.jpl.nasa.gov/.../sepo/education/nav/ss2.gif>

four planets as fingers. Jupiter is the other thumb with the outer set of four planets as fingers. Jupiter and Saturn are the biggest. Uranus and Neptune are medium sized. Venus, Earth and Mars are small. Mercury is tiny. Pluto discovered in 1930 was de-listed as a planet on 2006.

Pay no attention to that black dot in front of Jupiter. It is probably Jupiter's largest moon, Ganymede.

There are a lot of properties that can be connected to each of the planets, such as, their distance from the sun, their diameter, how many days it takes for them to orbit the sun, etc. However, like everything else in life, some things are more important than others. Mars and Pluto have little influence.

Sun And Planets	Distance From Sun Earth=1	Mass Ratio Earth =1	Magnetic Field Ratio Earth=1	Angular Momentum Ratio Earth=1	Tidal Force M/D3	Grav Rela M/D2	COM Disp MD
Sun				41			
Mercury	0.3871	0.0553	0.0006	0.0337	.97	.37	0
Venus	0.7233	0.815	0	0.0067	2.19	1.58	1
Earth	1	1	1	1	1.01	1.01	1
Mars	1.5237	0.107	0	0.13	.03	.05	0
Jupiter	5.0233	317.83	**19519**	700	2.26	11.76	1657
Saturn	9.5162	95.159	**578**	289	.11	1.05	908
Uranus	19.165	14.536	48	63	.00	.04	280
Neptune	30.003	17.147	27	93	.00	.02	520
Pluto	39.503	0.0021	0	0.0133			

Fig 13d: Planet Properties

Jupiter

Everyone knows Jupiter is the big kid on the block. But did you know, Jupiter has a magnetic field intensity 19,519 times that of Earth and over 29 times that of all the other planets, combined. Further, Jupiter has an angular momentum ratio of 700 compared to the Earth, 17 times higher than the Sun, and 1.43 times greater than all the other planets and Sun, combined.

Jupiter's atmosphere consists of clouds methane and ammonia gas. The structure of Jupiter's interior, like all the outer planets, is entirely speculative, since no exploration has been performed and no physical examination is possible. Only one spacecraft has penetrated any Jovian

Fig 13e: Jupiter Cross Section
www.planetsalive.com/image/jupxsec.jpg

atmosphere. On September 21, 2003, Galileo was deliberately destroyed, plunged into Jupiter's crushing atmosphere. This ended a 14 year odyssey that began with its launch on October 18, 1989 by Space Shuttle mission STS-34.

The upper layer of Jupiter is about 10,000 miles thick and consists of hydrogen and helium gas. Below this and to within 15,000 miles of the center of the planet is helium and **liquid metal hydrogen**. Jupiter is believed to have a rocky core about the size of the Earth, covered with about 10,000 miles of liquid ices. In short, Jupiter is a giant superconducting magnet.

Saturn

Fig 13f: Saturn
(Photo courtesy of NASA)

The internal structure of Saturn is similar to that of Jupiter. The icy/rocky core is larger, but the metallic hydrogen layer is about one-third the thickness.

Uranus

Uranus has the distinction of being the coldest of the four Jovian planets, even though it is closer to the Sun than Neptune by a billion miles. Unlike the other three Jovian planets, Uranus does not have a hot core, and radiates less heat than it absorbs.

Uranus is unique among the other eight planets in that it rotates almost on its side. More unusual, its magnetic axis is tilted 59 degrees away from its rotation axis. Even weirder, its magnetic axis does not pass through the planet's center, but is offset about 1/3 of the planet's radius.

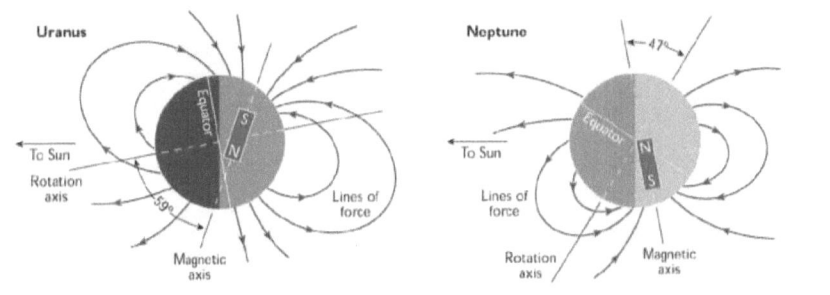

Fig 13g: Uranus/Neptune Axes as measured by NASA Voyager 2
<http://www.cambridgeblog.org/2018/01/the-magnetic-fields-of-uranus-and-neptune/>

Neptune

Neptune has a magnetic field 27 times that of Earth, and has an angular momentum 93 times that of Earth. Surprisingly, Neptune, even though it is 1 billion miles further away from the Sun, has a larger angular momentum than Uranus by a factor of 1.48.

Conclusion

What do all these details about the planets have to do with anything? With respect to how the Sun behaves and has an effect on the Earth's weather, the answer is: everything. As large as the Sun is, the fact is the planets have a profound effect on the Suns "weather" which in turn has a profound effect on the Earth's weather.

*"If automobiles had followed the same development **cycle** as the computer, a Rolls-Royce would today cost $100, get a million miles per gallon, and explode once a year, killing everyone inside."*

— Robert Cringely

Chapter 14: Solar Cycles

Our Sun goes thru a mind numbing quantity of cycles. Some as short as minutes, others that are over millenia long. By far the most notable of solar cycles is the Sun Spot cycle of 11 years. However, there are many others.

The 5 Minute Oscillation of the Solar Surface

In 1962 it was discovered by Leighton, Noyes and Simon that the lower layers of the Sun's photosphere contain standing acoustic waves (Ulrich, Roger K., The Astrophysical Journal, 162: 993-1002, December 1970), and that the energy dissipated by these waves was responsible for heating the chromosphere and corona. First, it is eye-opening to find out that cycles of such short duration could occur in a mass as large as the Sun. Second, it is intriguing that the hottest part of the Sun, it's chromosphere, operates many times hotter than the surface of the Sun, and the mechanism making this possible is acoustic waves. How could sound waves act as a temperature amplifier?

160 Minute Solar Cycle

In 1976, the science of helioseismology was born with the papers published by Brookes, Isaak and Raay and others (Royal Astronomical Society, 1978, **184**, 759-767). In this work it was determined that the surface of the Sun oscillates with a period of 160 minutes. Not surprisingly, this same 2.65 hour period cycle has been observed in the Earth's magnetic field.

154 Day Solar Cycle

In 1984, it was first discovered that gamma ray flares from the Sun follow a 154 day cycle (Rieger et al, Nature, 312, 623, 1984). Within the next ten years papers by Bogart, Bai, Lean and others confirmed the existence of a 154 day oscillation in the Sun's output of proton flares, solar wind, neutrinos and radio flux.

1.3 Year Solar Wind Oscillation

The IMP-8 and Voyager-2 spacecraft have discovered a strong modulation in the solar wind speed of about 1.3 years in duration. (Richardson, John d., et al, Center for Space Research, MIT Cambridge)

5.5 Year Hellman Cycle

In 1906, the Helmann Cycle was reported by the Prussian Meteorological Bureau. This half sun spot cycle was found to correspond with rainfall cycles in California and growth patterns of trees in Arizona. (Douglas, A. E., Evidence of Cycles in Tree Ring Records, PNAS, 1933 March, 19(3):350-360). Kocharov et al found the quasi five year variation in the abundance of nitrate in central Greenland ice.

1.1, 1.2, 2.8, 4.8 and 156 Month X-Ray Burst Frequency Cycles

Theodor Landscheidt found between 1970 and 1982 that X-Ray bursts occurred in 1.1, 12, 2.8, 4.8 and 156 month intervals. Dr. Theodore Landscheidt of the Schroeder Institute for Research in Cycles of Solar

Activity. He has been described as a modern Kepler. He was one of the first scientists to predict the arrival of El Ninos, using his groundbreaking research into solar cycles.

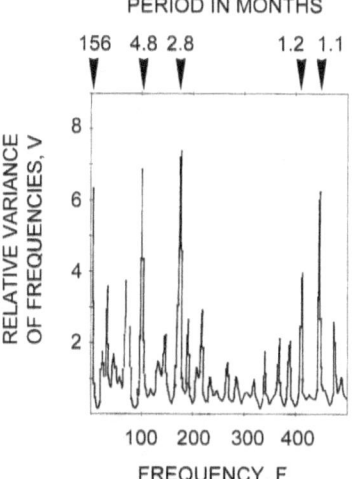

Fig 14a: Maximum Entropy Spectrum of X-Ray Bursts 1970 to 1982
(Landscheidt T., Sun-Earth-Man, Urania, 1989)

9-year, 2.25-year and 3 month Solar Cycles

Solar flares are related to weather, including atmospheric circulation changes, rainfall, and thunderstorm incidence. These weather patterns fall into 9-year, 2.25-year and 3-month cycles. <http://landscheidt. auditblogs.com/papers-by-dr-theodor-landscheidt/cycles-of-solar-flares and weather>

The 11-Year Schwabe Cycle

What if there was one mathematical expression that could summarize everything we have covered so far. There is. This mathematical expression is 11.1928695 * 2 ^ n, or 11.1928695 times 2 raised to the power of n, where n is the integer from -6 to +7. Sound preposterous? Let's have a look.

The basic solar sun spot or Schwabe cycle is about 11 years and occurs when n=0. When n=1, the result is the 22 year Hale magnetic cycle of the sun. When n=3, the result is 45 years. Wouldn't it be great if all the influences described above could be bound together in a coherent package explaining how everything fits together and why the climate changes? Well there is.

The most basic solar cycle is that of the sun spots and how they vary to a very predictable rhythm of 11 years, or so. This cycle was discovered by Heinrich Schwabe who documented his observations between 1826 and 1843. There have been 23 of these sunspot cycles completed thus far. Solar Cycle 1 started in 1755 and ended in 1766. Solar cycle 23 ended and Solar Cycle 24 began on January 2010.

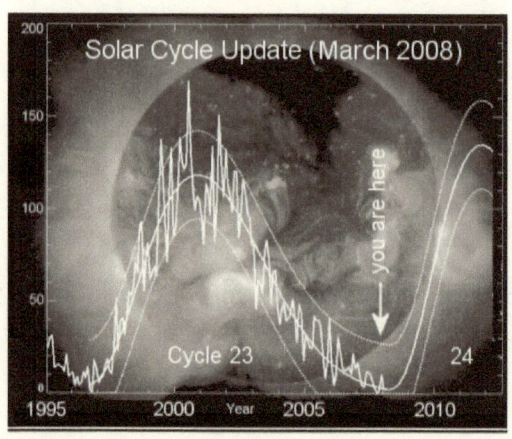

Fig 14b: Solar Cycle 23
<http://science.nasa.gov/science-news/science-at-nasa/2008/28mar_oldcycle/>

The consensus of scientists in March 2008 before the start of Cycle 24 was that it would be more active than Cycle 23 as depicted in the figure above. Well, the consensus looked around the classroom to see how many other hands were raised before it registered its vote and as usual was wrong. At this writing, NASA and NOAA have finally joined the growing number of scientists that Cycle 24 is going to be much less active than Cycle 23. This change in heart only came from the realization that actual data was proving beliefs to be wrong.

Any sunspot cycle has a characteristic shape where it starts out a little faster than it finishes. This kind of shape is very close to a Raleigh or Weibull Distribution Curve that is used in many analyses such as wind turbine analysis for how much power is available historically from the wind at any particular geographic location.

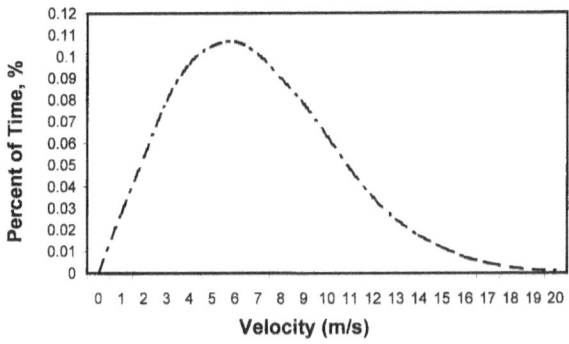

Fig 14c: Typical Wind Histogram

This shape has also been likened to follow the proportions of the Golden Section. The Greeks originally used this principle of proportion in designing their temples. The column height was considered the Major Height and was 61.8% of the overall height. The Minor Height of the entablature over the columns was the Minor Height, which was the

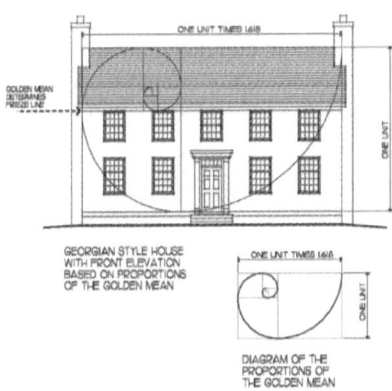

Fig 14d: The Golden Section
<www.pinterest.com/pin/322500023296930896/ >

remaining 38.2% of the overall height. What the Greeks had stumbled upon in their quest for a most pleasing design was the basic building block of cyclical structure. A most interesting property of these two numbers, 0.618 and 0.382, is that they are universal. These numbers are derived from the Golden Ratio, which is the ratio of the sum of the quantities to the larger quantity is equal to the ratio of the larger quantity to the smaller one. There is only one number that produces this result, it is the irrational number Phi (Φ) of 1.6180339887. Φ = a / b = a + b / a.

The Golden Section is prevalent in aesthetics, architecture, painting, book design, perception studies, music, industrial design, and finance. The Golden Section is the basis of the Fibonacci Spiral so often seen in nature. The omnipotence of this principle is far-reaching and breathtaking.

But this simple Rayleigh- looking curve often has a double peak. It has been deduced by K. Geogieva of the Space and Solar-Terrestrial Research Institute of the Bulgarian Academy of Science that this double peak is caused by a manifestation of the two surges of the toroidal field of the Sun <arxiv.org/pdf/1103.45552>.

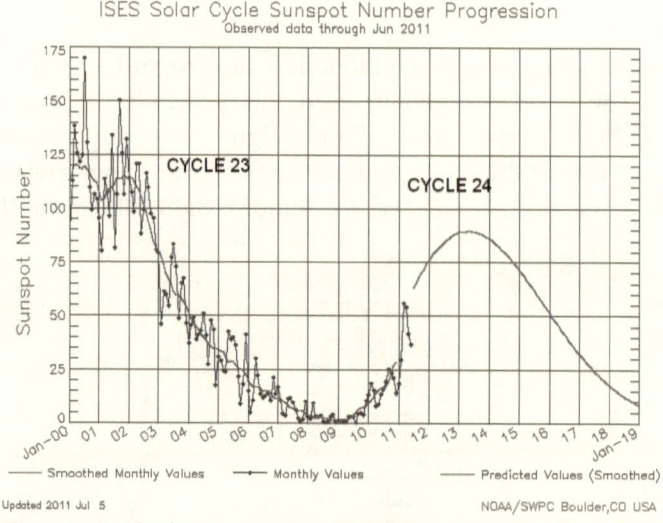

Fig 14e: End of Solar Cycle 23 and Start of Solar Cycle 24
<http://www.swpc.noaa.gov/SolarCycle/>

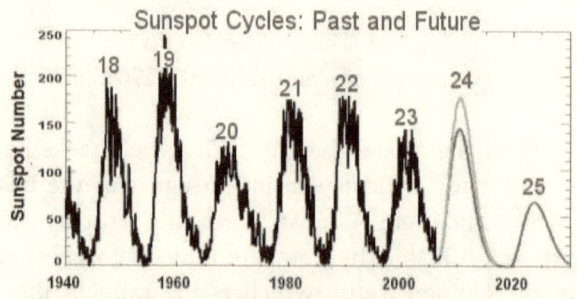

Fig 14f: Sunspot Cycles 18 through 25
http://science.nasa.gov/science-news/science-at-
nasa/2006/10may_longrange/

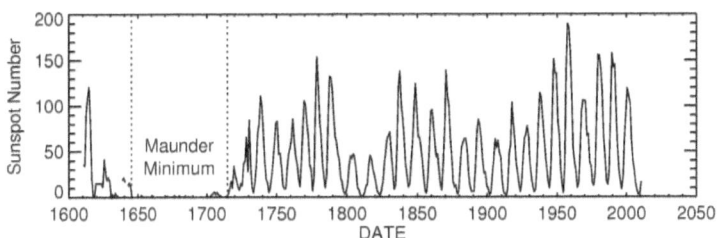

Fig 14g: Schwabe Cycle of Sun Spots
(Credit NASA/MSFC)

Looking at sun spot cycles in themselves is revealing. But oftentimes looking at rates of change of this data yields even more insights. Take for example the Sun's angular momentum. Of itself it looks mundane. But taking the directive of this change reveals something almost uncanny.

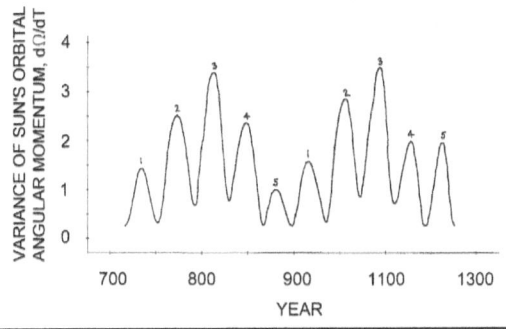

Fig 14h: Variation in Sun's Orbital Angular Momentum
<http://plasmaresources.com/ozwx/landscheidt/pdf/
Solar_Activity_A_Dominant_Factor_In_Climate_Dynamics.pdf>

The figure above shows the rate of change of the Sun's orbital angular momentum. The curve has been described as looking like "two hands" forming the Solar System, with the Sun and Jupiter being the thumbs and the other eight planets being the fingers. The period or spacing between fingers is 35.6 years on average. The five finger sequence is 178.8 years, just twice that of the Gleissberg cycle.

The 13.3-Year Cycle

A 13.3 year cycle has been derived from spectral analysis of surface temperature, atmospheric pressure and zonal westerlies. (Rampino et al, Climate – History, Periodicity, Predictability, Van Nostrand Reinhold, New York NY, 1987) (page 421).

Ellery Hale discovered in 1925 that the magnetic polarity of sun spots changed with each 11-year sun spot cycle. Thus, every 22 years the polarity of the sunspots made a complete cycle from north to south and back. Scientists use this marker at the end of one sunspot cycle as the number of sun spots wind down, and look for a change in polarity as the confirmation that the next sunspot cycle has begun.

This happened much slower between the old cycle 23 and the present new cycle 24. Cycle 24 was about 2.5 years late.

The 31-year Sahelian Drought Cycle

The Sahel is the area in Africa between the Sahara to the north and the Sudan savanna to the south, running from the Atlantic Ocean and the Red Sea. In this area there is a 31-year cycle, thought to be a cross correlation of the 11-year Schwabe and 22-year Hale cycles. This was first noticed by Sir Francis Bacon in 1625 and later by Eduard Bruckner. Rhodes Fairbridge also recognized a complicated relationship between Nile floods and Sahelian droughts.

The 35-year Bruckner Cycle

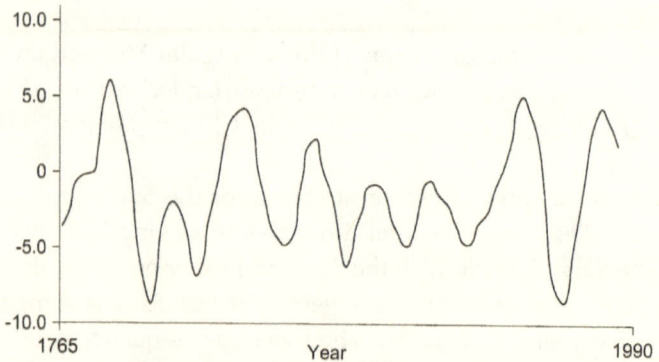

Fig 14i: The 35-year Bruckner Cycle
<http://garymorris93.cwahi.net/weather/solar_variations.html>

Eduard Bruckner proposed a 35-year cycle which appeared in an article The Southern Ocean. It's Climatic Control Over Australia. Address by Professor Gregory Wellington, Tuesday in The Argus, Melbourne, Victoria, Australia, Wednesday January 13, 1904,.

<https://trove.nla.gov.au/newspaper/article/10592456?searchTerm=antarctic%20ice%20melting&searchLimits=>

Much of the material that Bruckner drew upon was from European sources including secular variations in the Caspian Sea, lakes and seas without outlets, river heights, precipitation, atmospheric pressure, occurrence of severe winters, and advance and retreat of glaciers. No computer projections here; just good old-fashioned data gathering and the drawing of fundamentally sound conclusions.

83-year Secular Cycle

A distinct period of decadal modulation of 11-year cycles has been acknowledged by science as a "secular" sunspot activity of about 80 years.

It has been determined from an analysis of 7,600 years of data, that there is a correlation with climate, volcanism and ozone column to a wave pattern formed by secular variations in the 'impulses of torque' driving the sun's oscillatory motion about the center of mass (COM) of the solar system. This "impulse of the torque" (IOT), or differential torque versus time (dT/dt) has been defined by Landscheidt (Climate, page 421). This period is 80 years.

The 88-year Gleissberg Cycle

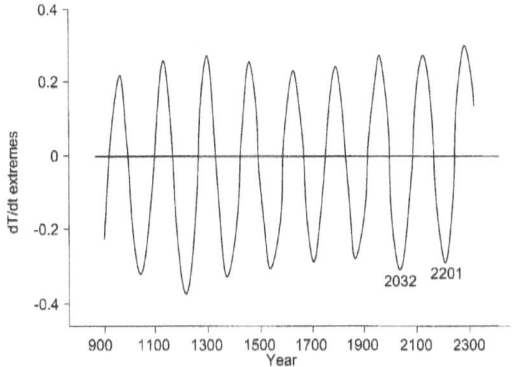

Fig 14j: Gleissberg Cycle of Changing Sun Spot Amplitude
(Based on Landscheidt , Climate, page 421)

Note above that the next extreme minimum will occur in the year 2032.

145

This is a cycle of changing amplitude of the 11-year sun spot cycles. The period of this cycle varies from about 80 years to 100 years, depending on how the data was reconstructed. Using 11,854 years of carbon-14 de-trended tree ring data it is 87.8 years. Using sunspot numbers from 1700 to 2005, it is 87 +/-13 years. Using modulation of the length and symmetry of the Schwabe cycle from 1853 to 2000, it is about 90 years. Using reconstructed solar irradiance from 1600 to 2000, it is about 80 years. Using solar proton events form 1561 to 1994, it is about 85 years. Using Ti levels in stony meteorites from 1750 to 2000, it is about 100 years. Using GISP2 dust profiles, it is about 91 years. Taking an average of the seven, it is about 88.7 years.

This cycle also shows up in global air and ocean temperatures suggesting a sun/climate connection. A secular cycle of solar activity is related to the sun's oscillatory motion about the center of mass of the solar system. This motion imparts relatively strong impulses of torque at a mean interval of 19.86 years. Four consecutive impulses define a permanent wave with a period of 79.46 years. (Landscheidt, T., Swinging Sun, 79-year cycle and climate change, J. interdiscipl. Cycle Res., Vol. 12, No. 1, 3-19, 1981)

The 120-Year Solar Cycle of the Cosmogenic Isotopes

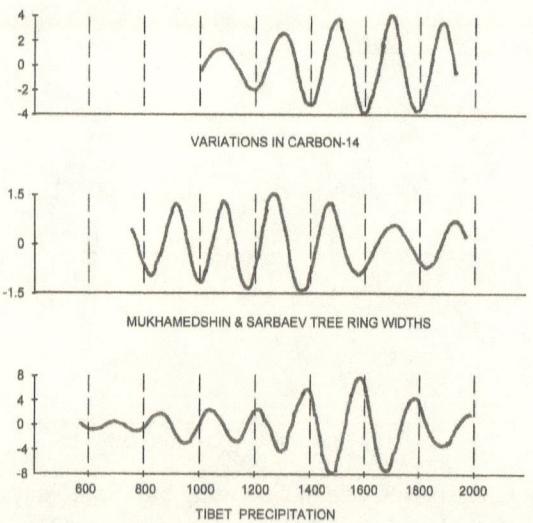

Fig 14k: The 120-Year Solar Cycle of Cosmogenic Isotopes
(Velesco VM et al, ICRC,1,553-556)

The figure above depicts 120-year cycle variations in Carbon 14, tree ring widths, and Tibet precipitation. Here is evidence of tree ring growth variations and precipitation variations marching to the beat of solar activity and cosmic ray flux.

Cosmogenic isotopes are radioactive fingerprints left on things on and in the Earth. The amount of isotopes deposited is a function of solar activity. The more active the sun, the more cosmic rays are allowed to penetrate the Earth's protective magnetosphere and irradiate elements on the ground. These isotopes can be Beryllium-10 (Be^{10}) which is found in glaciers, or Carbon-14 (C^{14}) which are left in tree rings.

208 Year deVries Cycle

The deVries Cycle is one of the more intense solar cycles. Depending on who is doing the writing, it is has been called the 200-year, 205-year, 207-year, 208-year, and 210-year deVries Cycle. They are all the same.

The period of this cycle varies from about 188 years to 210 years, depending on how the data was reconstructed. Using 11,854 years of carbon-14 de-trended tree ring data it is 208 years. Using sunspot numbers from 1700 to 2005, it is 188 years. Using Beryllium-10 levels in GRIP ice core between 25,000 and 50,000 years BP, it is about 205. Using GISP2 dust profiles, it is about 197 years. Taking an average of the four, it is about 199.5 years.

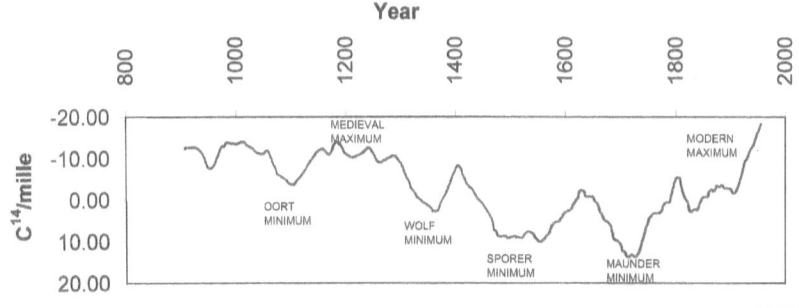

Fig 14l: The (205 Year) deVries Cycle
< http://en.wikipedia.org/wiki/File:Carbon14_with_activity_labels.svg>

What is interesting to note about the graph above is that the vertical axis is not temperature, but carbon-14 measurements. If it were temperature, AGW advocates could make the claim of human induced climate change.

But these measurements are natural and come from solar activity, not man made activity.

This cycle records periods of deep solar inactivity. During these periods the Earth experiences profoundly colder weather. Previous deVries cycles counting back from the present include the Dalton Minimum between 1790 and 1810, the Maunder Minimum between 1645 and 1715, the Spörer Minimum between 1450 and 1550, the Wolf Minimum between 1280 and 1350, the Medieval Minimum between 1110 and 1250, and Oort Minimum between 1040 and 1080. Twenty three other Grand Minima have been identified from 690 AD to 9170 BC. (Usoskin et al, 2007, "Grand minima and maxima of solar activity: new observational constraints", Astron, Astrophys, **471**: 301–9).

In the graph below, these deVries Cycles are characterized by the roughly 200 year peaks occurring at years 1000, 1200, 1400, 1600, 1800 and 2000.

The 314-year Elatina Cycle

This cycle is derived from 52, 63 and 105 years. The triple conjunction of Jupiter, Saturn and Uranus occurs every 317.7 years. (Landscheidt, p79)

There is significant correlation with the conclusion of the above and a cycle called the Elatina Cycle. The Elatina Cycle gets its name from the layers or varve found in a core boring of 10 meters in depth made in a lake deposit in the Flanders Ranges of South Australia. (Williams, G. E., Solar Affinity of Sedimentary Cycles in the Late Precambrian Elatina Formation, Aust. J. Phys., 1985, 1027-43). The analysis of this 33 foot deep core reveals a sequence of 1,580 cycles, each with about a 12-year cycle, for a total of 18,960 years of data. In addition, for about every 26 cycles there is a maximum thickness that occurs. This maximum varies between 276 and 336 years, with a mean centered at 314 years. That equates to 60 of these 314 year cycles. Does this correlate with the five 308-year cycles of Kouwenberg data?

The 1470-year Bond Cycle

The Bond Cycle is named after Dr. Gerard C. Bond who isolated eight such cycles in the current Interglacial warmup period. The evidence of these cycles is registered as ice rafting debris (IRD) deposited onto the ocean floor carried by ice bergs from the North Atlantic. When there are few ice bergs the deposits in core sample layers are nil. When there are lots of ice bergs, there are significant deposits of earth in core sample layers. <https://www.britannica.com/science/Bond-cycle#:~:text=Bond%20event%2C%20also%20called%20Bond%20cycle%2C%20any%20of,years%20ago%20and%20extending%20to%20the%20present%20day%29>

The 2,400-year Hallstadt Cycle

The 2,400-year Hallstadt Cycle is yet another cosmogenic isotope cycle of solar activity, but of greater length. Here, the 120-year cycle is absorbed and what appears in its place is an extended cycle of about 20 times the length, or 2,400 years. <wattsupwiththat.com/2016/11/24/the-bray-hallstatt-cycle/>

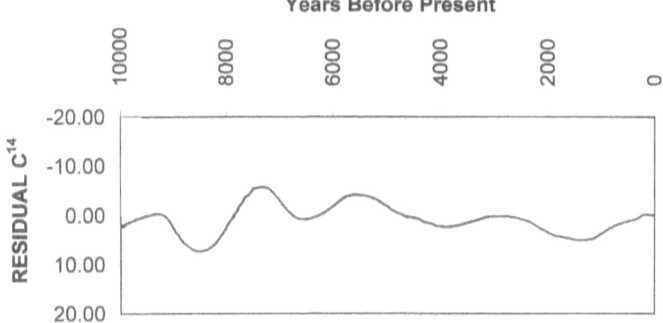

Fig 14m: The 2,400-Year Hallstadt Cycle
< http://en.wikipedia.org/wiki/File:Carbon-14-10kyr-Hallstadtzeit_Cycles.png>

The 2,402-year Charvatova Cycle

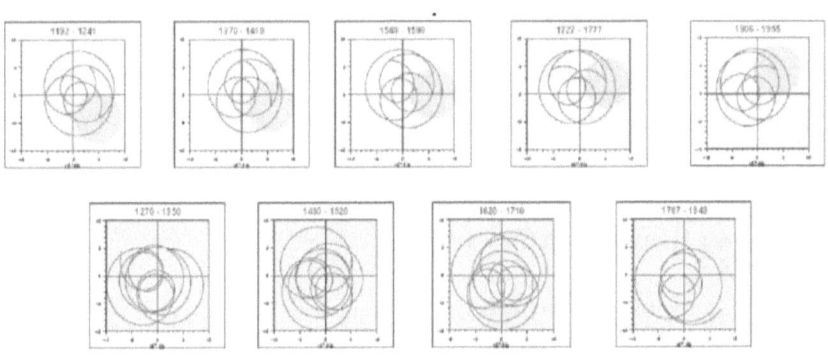

Fig 14n: 370-year segments of Exceptional, Stable Pattern of SIM

149

Ivanka Charvatova's field of study is solar inertial motion (SIM). According to Charvatova SIM can be classified into two elementary types: a trefoil-like trajectory governed by the Jupiter-Saturn order and a chaotic motion. The first lasts for over 50 years and occurs every 179 years (J-S). The 2,402 year period is governed by Uranus/Neptune (UN).

The 6,000-year Cycle

Over a period of 11,360 years a long term 6,000 year cycle has been discovered among all the 11-year sun spot data. (Xapsos, M. A. and Burke, E. A, Solar Physics, Evidence of 6,000-Year Periodicity in Reconstructed Sunspot Numbers, Vol. 257, No. 2, 363-369, 2009)

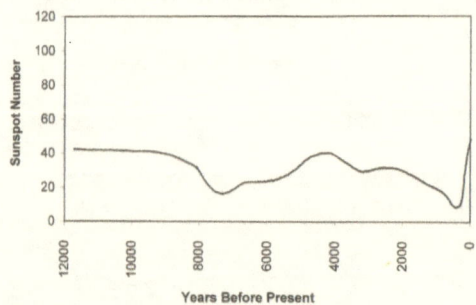

Fig 14o: The 12,000-Year Sun-Spot Number Cycle
<http://en.wikipedia.org/wiki/File:Sunspots_11000_years.svg>

The number of sunspots in any given year is a barometer of solar activity. The more activity, the colder the climate usually becomes. As can be seen above, the nadir of sunspot activity occurred about 400 years ago, which resulted in the Little Ice Age. Look at the slope, or rate of change, of the SSN within the last couple hundred years to the right on the graph above. The SSN has recently skyrocketed. Talk about Hockey Sticks!

Questions

- Is there a correlation between the 314-year Elatina Cycle and 308-year Kouwenberg CO2 cycle data?

- Are Jupiter-Saturn and Uranus-Neptune the major players in the controlling the intermediate and long term cycles of the Sun?

*"Normal is nothing more than a **cycle** on a washing machine."*

– Whoopi Goldberg

Chapter 15: Lunar Cycles

There is nothing closer to Earth, or has more proximate influence on our climate and us, than the Moon. To ignore the moon ignores the obvious. It is the pacemaker of animal rhythms, such as, menstrual cycles. It tugs away at the oceans changing their levels by about 6 feet every 5-1/2 or so hours. How can this enormous influence be so casually overlooked by warmists?

<u>Temporal, Measured or Calendar Cycles</u>

There are numerous cycles associated with the moon which have developed over thousands of years. These are of the measurement variety; they are useful to quantify rotations or orbits, but otherwise do not have any apparent impact on us or the Earth.

The Saros Cycle has its origins in ancient Babylonia. The name was coined by Edmond Halley in 1691. Saros comes from the Babylonian word "Sar", which was a unit of measure with a number of 3600. This is the eclipse cycle of 18 years + 11 days + 8 hours. This is the sidereal period for the moon to complete a revolution relative to the fixed stars of 27.322 days. This is the synodic period of the moon to complete a revolution relative to the Earth-Sun line. This is the period for lunar phases as seen from Earth called 'a month', or 29.53 days.

The Metonic Cycle was discovered by Meton, a Greek mathematician, astronomer and engineer in 432 BC and reported in his book Enneadecaterides. Meton noticed that 19 solar years was exceedingly close to 235 moons. He saw that 19 solar years and 235 lunar months both added up to 6,940 days. The ratio of 235 moons to 19 years was found to be embodied in the gear train of the Antikythera mechanism, an astronomical computer probably built before 80 BC and recovered in 1900 from a shipwreck off the Greek Island of Antikythera.

The above cycles are more for curiosity than having to do with climate. Nevertheless, cultures developed these cycles because they had an influence on their lives. Then there are cycles which have been discovered that do have an impact on us, our weather and our climate.

The 1,800-year oceanic tidal cycle

Ice rafting debris layers as discovered in sedimentary core records reveal an 1800-yr tidal cycle. This strong tidal forcing causes cooling at the sea surface by increasing vertical mixing on the oceans. An amplitude modulation of this cycle occurs about every 5,000 years. (Keeling and Whorf, The 1,800-year oceanic tidal cycle: A possible cause of rapid climate change, Vol.97, Issue 8, 3814-3819, 2000). Dr. James Totts Hansen, please take note. These were abrupt climate changes that occurred well before any fossil fuel burning.

The 18.6-Year Lunar Nodal Cycle

An 18.6-yr lunar nodal cycle has been discovered as a significant feature of winter air and sea temperatures along the North American west coast over a 400-yr period. (McKinnell and Crawford, The 18.6-year lunar nodal cycle and surface temperature variability in the northeast Pacific, JGR, **112**, 2007)

There has been a lot written recently in the MSM about Arctic Sea Ice Extent, forecasts of the disappearance of the Arctic Ice Cap, and so on. These ramblings are not new, and have been going on since the 1960's.

An 18.6-year, and 74-year sub-harmonic, has been identified which influences long-term tides, polar motions, Arctic Ice Extent, the NAO winter index, weather and climate (Yndestad, Harald, The influence of the lunar nodal cycle on Arctic climate, ICES Journal of Marine Science, **63**, 401-420, 2006). Al Gore and the IPCC please take note! Changing arctic sea ice extent has nothing to do with CO_2, and everything to do with the Moon (ie: celestial mechanics).

18.6-Year Declination Cycle

A link has been found to sea level changes and the 18.6-year oscillation of the Moon's orbital declination (Denny, M. W. and Paine, R. T., Celestial Mechanics, Sea-Level Change, and Intertidal Ecology, The Biological Bulletin, Vol. 194, Issue 2, 108-115, 1998).

When the declination is at its minimum of 18-deg, extremes in weather including heat waves, record snow falls, blizzards, floods, cyclones, and gales are prevalent in the Southern Hemisphere. When the declination is at its 28-deg maximum long droughts can be expected. When the declination is at its 23.5-deg midpoint, average conditions return and El Ninos are common in the Southern Hemisphere. The opposite conditions can be expected in the Northern Hemisphere

The 9.3 year Lunar Half-Nodal Cycle

A 9.3-year Lunar Half-Nodal Cycle has been found, which gives a good fit to the peak abundance of Snowshoe Hare, Ruffed Grouse and Canada Lnyx. The years of abundance follow the formula $P = 1950.22 +/- 9.3n$. The vernal equinox of 1950 represents the point of the moon's maximum declination. (Archibald, Herbert L., Wildlife Society Bulletin, Vol. 5, No. 3, 1977). This 9.3-year cycle is a harmonic of the 18.6 year lunar nodal cycle.

4.425 Perigee Cycle

The 4.425 year Perigee Cycle describes the changing distance between the Earth and Moon. A surprising number of varve (a pair of alternating course and fine layers believed to comprise an annual deposition in a body of still water) and total organic carbon cycles can be interpreted as deriving from the 18.6-year nodal and 4.425-year perigee cycles of the moon. (Berger et al, Tidal cycles in sediments of Santa Barbara Basin, Geology, Vol. 32, No. 4, 329-332, 2004)

Daily Cycles

The gravitational effect of the moon as the Earth turns daily has a profound effect of changing the ocean's level with the Neap and Ebb Tides. This phenomenon has not gone unnoticed by advocates of tidal energy generation. So, why would it be a surprise to anyone?

The world's oceans are yanked up and down twice a day. Although, our bodies are 70% water, we do not change in elevation as with the ocean's tides, or are consciously aware of the moon's attraction. However, there is no doubt the Moon affects us in many different, subtle ways. So, why would not plants notice this effect since they too are mostly water?

Volcanic Eruptions

A correlation has been documented showing a spike in volcanic activity of Stromboli in the Aeolian Islands of Italy at a point in time just between the full moon and perigee (Handwerk, Brian, Are Volcanic Eruptions Tied To Lunar Cycles, National Geographic News, February 15, 2002).

U. S. Rainfall Maxima

Every 14.6 days one of the 29 high tides that occur during this period reach their highest point. A correlation has been found with these peak tides and the amount of rainfall in the US. Within cycles from full moon to full moon and from new moon to new moon, rainfall maxima coincide with the major (0.618) of the golden section, whereas rainfall minima coincide with the minor golden section (0.382) (Landscheidt, T., The Golden Section: A Building Block of Cyclic Structure, Paper presented at Conference, 1992).

Other Phenomena

About 96% of Earthquakes occur on, or within a day of, the new Moon, Full Moon, or Perigee. (Ring, Ken, Predicting the weather by the moon, Gothic Image Publications, Glastonberry, England, 2000).

A daily global temperature change of 0.02K has been discovered between the new moon and full moon (Balling and Cervany, Influence of Lunar Phase on Daily Global Temperature, Science, Vol. 267, No. 5203, 1481-1483, 1995). With light coming from the moon there must be heat. It just couldn't be that much. Now you know how much it is.

More unintentional poisonings have been called in to the Maryland Poison Control Center during 13 lunar cycles (Oderda and Klein-Schwartz, Lunar Cycles and Poison Center Calls, Informa Healthcare, Vol. 20, No. 5, 487-495, 1983).

A correspondence has been discovered between lunar cycles and emergency room visits as reported by the New England Journal of Medicine <www.nejm.org/doi/full/10.1056/NEJM197806082982318>.

Unnoticed by humans because they are too small, are lunar induced winds, which reach 1/20 of a mile per hour. In the morning they move east, in the evening they move west. (Crawford, E. A., The Lunar Garden: Planting by the Moon Phases, Capital Books, 2000)

Peter Yankley of the University of New Orleans has found evidence that the phases of the moon drive hurricane behavior (Reilly, Michael, Lunar Cycle Turns Hurricanes Into Beasts, Discovery News, March 5, 2009).

University of Michigan Professors Dichev and Janes have found that stock market returns in the 15 days around the new moon dates are about double the returns in the 15 days around the full moon dates (Walker, Tom, Article: Study Shows correlation between stock market, lunar cycle, Atlanta Journal Constitution, November 29, 2008)

A link has been found between sperm whales being stranded and lunar cycles (Wright, A. J. (2005), lunar cycles and sperm whale (physter macrocephalus) strandings on the north Atlantic coastlines of the British Isles and Eastern Canada, Marine Mammal Science, 21: 145–149. doi: 10.1111/j.1748-7692.2005.tb01214.x). Perhaps this study will quell any notions about whales being stranded caused by global warming.

Summary

- The 1,800-year lunar declination cycle corresponds to an 1,800-year oceanic tidal cycle, which causes cooling of sea surface temperatures with an attendant 5,000-year amplitude cycle.

- An 18.6-year, and 74-year sub-harmonic, has been identified which influences long-term tides, polar motions, Arctic Ice Extent, the NAO winter index, weather and climate.

- The 9.3-year lunar half-nodal cycle is correlated with dozens of other cycles of wildlife and plant abundance.

- A link has been found to sea level changes and the 18.6-year oscillation of the Moon's orbital inclination.

- A 9.3-year Lunar Half-Nodal Cycle has been found, which gives a good fit to the peak abundance of three wildlife species.

- The 4.425 year Perigee Cycle is correlated with alternating course and fine layers of sea sediments.

- A correlation has been found with one of every 29 high tides with the amount of rainfall in the US.

- A correlation has been documented showing a spike in volcanic activity between the full moon and perigee.

- A daily global temperature change of 0.02K has been discovered between the new moon and full moon.

- The phases of the moon have been found to drive hurricane behavior.

*"If you want to find
the secrets of the
universe, think in
terms of energy,
frequency and
vibration."
— Nikola Tesla*

Chapter 16:
Earth Cycles

Hopefully, everyone knows the Earth rotates on its own axis about once every 24 hours. Besides this motion, the Earth has three other primary motions, including precession, tilt, and eccentricity. These motions have been called the Milankovich Cycles. The Earth also wobbles ever so slightly.

Precession of the Earth's axis is best described by comparing it to the motion of a spinning top as it slows down and begins to wobble wildly. The Earth's precession cycle occurs roughly every 23,000 years.

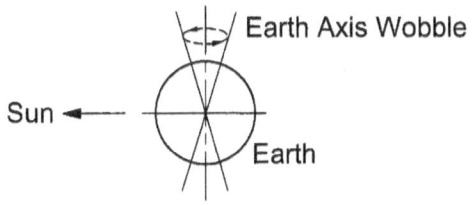

Fig 16a: Earth's Milankovich Precession Cycle

It is believed precession is due to the Earth not being perfectly round and having a bulge due to tidal forces. The resulting wobble causes the equinoxes and solstices to vary about the earth's axis.

Tilt is the angle at which the Earth rotates with respect to the Sun. This angle of tilt varies between 21.5 and 24.5 degrees. The period for this angle of tilt to complete one cycle is about 41,000 years.

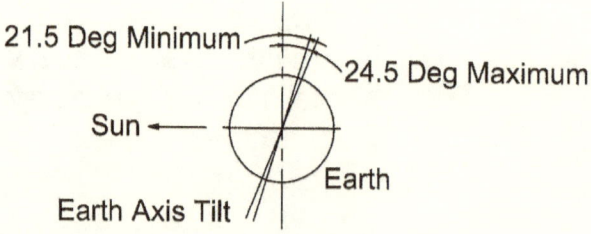

Fig 16b: Earth's Milankovich Tilt Cycle

As the Earth tilts towards the Sun, the North Pole is exposed to more sunlight. This annual warmup causes a reduction in the Arctic Ice Cap extent from about 14,000,000 km2 to about 6,000,000 km2. This accounts for he expression of northern countries being in the land of the midnight Sun; the Sun never sets in the summertime.

Eccentricity is the variation of the Earth's orbit around the Sun from a circle to an ellipse. This cycle takes about 100,000 years.

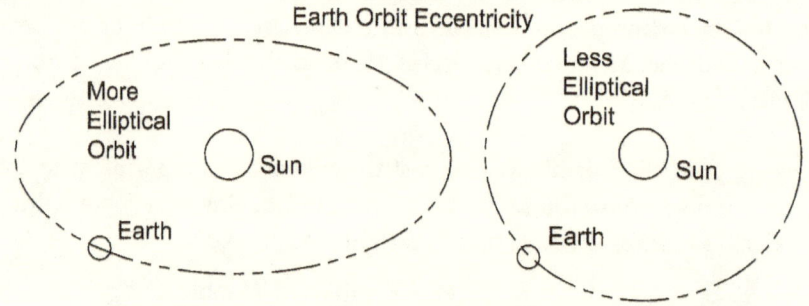

Fig 16c: Earth's Milankovich Elliptical Cycle

There is strong evidence that the occurrence of Ice Ages in 100,000 year cycles are primarily due to the Earth's Milankovich Eccentricity Cycle. This Ice Age cycle has been found to lag temperature, CO2 and the orbital eccentricity cycle (Shackleton, Nicholas J., The 100,000-Year Ice

Age Cycle Identified and Found to Lag Temperature, Carbon Dioxide, and orbital Eccentricity, Science, Vol. 29, No. 5486, 2000).

Polar Wander is the change in position of the Earth's rotation axis with respect to the earth's surface. Variations in sea level, the rise and decline of Ice Ages, and global change on geological time scales have been observed to coincide with polar wander. (Vermeersen and Sabadini, Polar Wander, Sea Level Variations and Ice Ages, Surveys in Geophysics, Vol. 20, No. 5, 415-440, 1999). IPCC and Al Gore, pay attention. Sea level variations can be due to something else besides melting glaciers. Do you think these findings are dialed into computer models? Not on your life.

Chandler's Wobble was discovered by Astronomer Seth Carlo Chandler in 1891. The displacement of the Earth's axis is about 20 feet and occurs about every 433 days. This wobble combines with another wobble with a period around 1 year, resulting in a polar motion of about 7 years. <http://en. wikipedia.org/wiki/Chandler_wobble>

On March 3, 2010 an 8.8 magnitude earthquake struck Chile. According to NASA JPL it was enough to change the way the Earth wobbles by about 3 inches. <http://www.npr.org/templates/story/story.php?storyId =124249439>

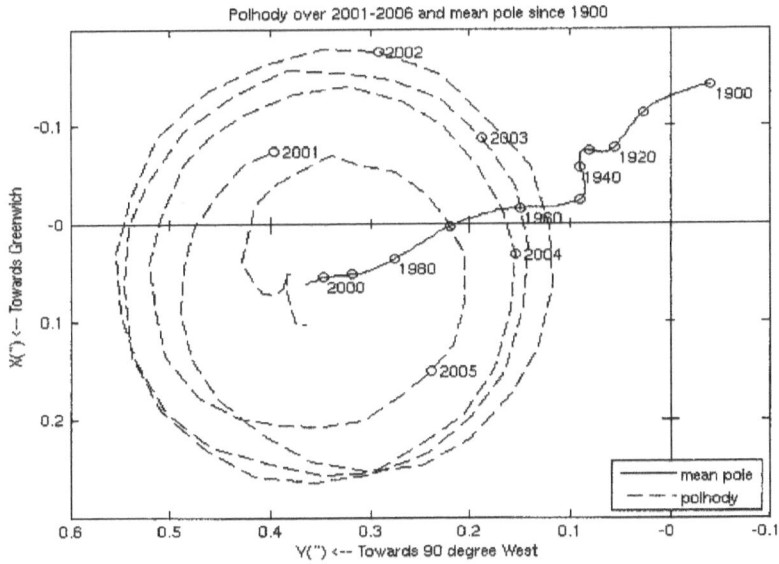

Fig 16d: Chandler's Wobble
<www.solarsystems.nasa.gov>

159

Note the plot above. The dotted "polhody" line shows a collapse of Chandler's wobble occurring from about 2005 to 2006. A polhode is the curve produced by the angular velocity vector on the inertia ellipsoid. (I guess that definition didn't help either). Just think of this path as the path the Earth's axis is taking about its axis if it did not move at all.

Two causes of Chandler's Wobble were identified (R. Gross, GRL August 1, 2019). Fluctuating pressure on the bottom of the ocean, caused by temperature and salinity changes and wind-driven changes in the circulation of the oceans.

Earth's Nutation

The question arises: "What causes the 60 year cycle?'

A further analysis of the Earth's wobbling suggests a residual motion of the polar axis. A plot of this motion is shown below

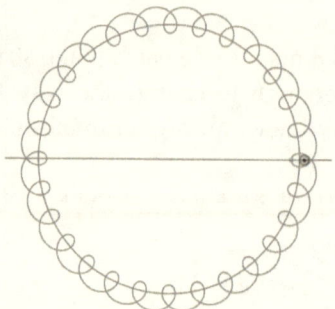

Fig 16e: Residual Motion of Earth's Polar Axis

The epitrochoid curve above is generated from two mathematical expressions. :

T1 =11.12*9.3* (11.12+9.3) = 5 years

T2 =11.12*9.3* (11.12-9.3) = 60 years

The 11.12 is the sunspot cycle. The 9.3 is one half of the lunar nodal cycle of 18.6 years. The T1 result of 5 years might suggest the origins of El Nino events, which spans regular 3 to 7 year cycles. The T2 result of 60 years suggests a correlation with the Atlantic Multidecadal Oscillation. And why should it be a surprise that the moon with it's cyclical gravitational pull and the Sun with it's cyclical irradiance be the cause of two of the most influential weather producers on our planet.

Shumann Resonance

All physical objects have a natural frequency. Tesla demonstrated this by almost bringing down an office building in Manhatten. Even the Earth has a natural frequency. It is about 7.3 cycles per second and has been increasing recently and unexplainably. Some scientists have claimed that the great Pyramid of Giza can harness the Earth's energy using the principle of resonance.

> *"In this respect, the history of science, like the history of all civilization, has gone through **cycles.***
>
> *– Abdus Salam*

Chapter 17: Planetary Cycles

The effect on our climate by the Sun is obvious. The effect of the Moon on our tides is also well known, though its effect on ocean circulations, which in turn does effect climate change, is not. Perhaps the best kept secret of all is the affect of the planets on our climate. At first this sounds preposterous. How can planets, so far away have any influence at all? Well, sit back and relax. You are about to be astounded by what happens when the planets interact and gang-up on the Sun.

Keplerian Mechanics

Picture a planet like Jupiter revolving around the Sun. To someone unfamiliar with Keplerian mechanics, or just simply forgets about action and reaction, the orbit of this planet most frequently envisioned is that of a circle.

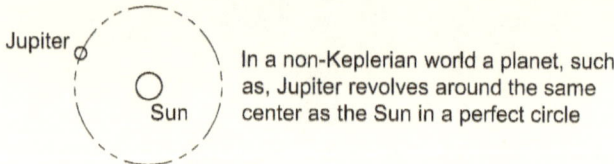

In a non-Keplerian world a planet, such as, Jupiter revolves around the same center as the Sun in a perfect circle

Fig 17a: Planet orbiting Sun in a Non-Keplerian World

However, this picture is not correct. The orbit must be an ellipse. Why? Because every action must have an equal and opposite reaction. It is not possible for the Sun to stay in one place with a planet tugging away at it as it rotates around the Sun. This is Kepler's First Law: The orbit of a planet is an ellipse, with the Sun at one focus of the ellipse and the planet at the other focus.

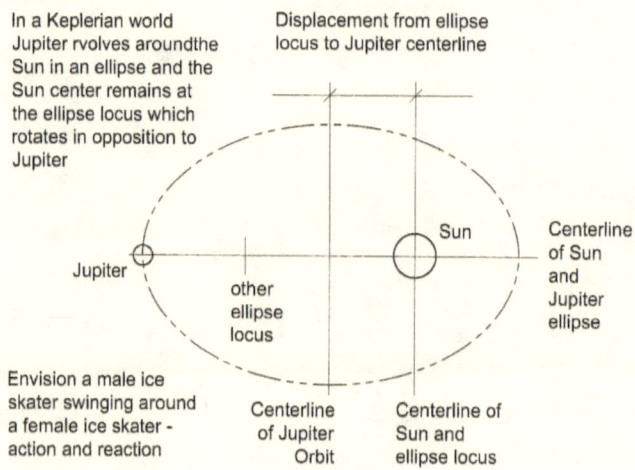

Fig 17b: Planet orbiting Sun in a Keplerian World

Pair Ice skaters demonstrate this principle more clearly. The male ice skater weighing 180 lbs spins around the spin axis with his center of

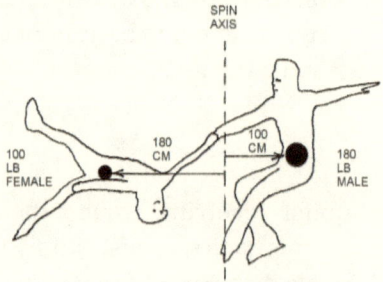

Fig 17c: Ice skating Pair

gravity at 100 cm from the spin axis. The female skater weighing only 100 lbs balances out the system with her center of gravity 180 cm from the spin axis. The forces must balance. If the skater's weight vary, the center of gravity distance must vary, inversely.

<u>The Solar System Barycenter</u>

Now picture two planets orbiting the Sun at different distances from the Sun. When the two planets line up on one side of the Sun, the tugging is amplified. For simplification orbits are shown as circles, not ellipses.

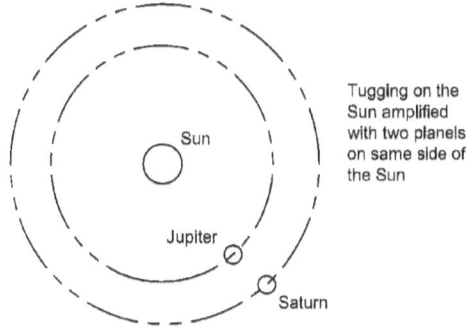

Fig 17d: Two Planets in Conjunction Around Sun

When the two planets are on opposite sides of the Sun, their tugging influence is somewhat neutralized. If the planets were of identical mass, their effects would be cancelled out.

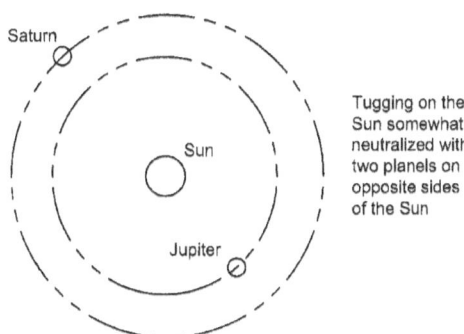

Fig 17e: Two Planets in Opposition Around Sun

For practical purposes it would at first appear that the effects of the small inner planets Mercury, Venus, Earth and Mars can all be eliminated. Based on their tiny masses and small angular momentum their effect on the Sun is negligible compared to the four huge Jovian planets Jupiter, Saturn, Uranus and Neptune. But as you will see shortly, Earth and Venus have a lot to say about what happens to the Sun.

Now picture all nine planets of the solar system rotating around the Sun. The effect is very hard to comprehend, but it is not impossible to illustrate. The Center Of Mass (COM) of all the planets is not fixed at the center of the Sun. As each planet rotates around the solar system center of mass of all the planets whirls about the Sun in an almost random spiral. This COM is the Barycenter of the Solar System.

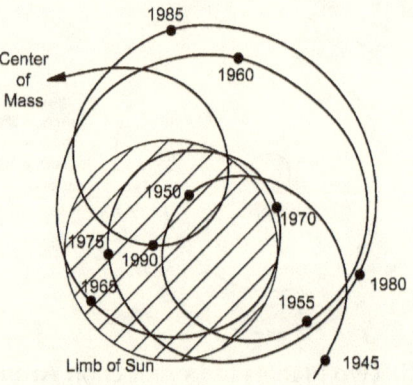

Fig 17f: Barycenter of the Sun
<http://wapedia.mobi/en/File:Solar_system_barycenter.svg>

The extent of the variation of the barycenter varies from 0.033 solar radii to 2.169 solar radii. The 2.169 is the sum of all planet radii. The 0.033 is the solar radii of Jupiter minus every other planet. This range is arrived at by the sum and subtraction of the data of the barycenter of the individual planets as follows:

Planet	Barycenter in Solar Radii
Mercury	0.000
Venus	0.000
Earth	0.001
Mars	0.000
Jupiter	1.068
Saturn	0.586
Uranus	0.180
Neptune	0.335
Pluto	0.000

*"One of the most
fascinating things about
golf is how it reflects
the **cycle** of life. No
matter what you shoot,
the next day you have to
go back to the first tee
and begin all over again
and make yourself into
something."*

- Peter Marshall

Chapter 18:
Coupled Cycles

The basic concept involves understanding how the planets influence the sun, and how the sun and moon influence the Earth. That's it. No CO_2, no melting ice, no polar bears, no cow farts, no SUVs, just the impact of celestial mechanics on the Earth.

Most people are intuitively aware that things do not happen instantaneously. All systems, be they man made or natural, have time lags built in. This is because all systems, to one degree or another, have inertia. When you step on the accelerator of your car it takes a fraction of a second for the mechanical linkage to activate the injectors, for the injectors to introduce more fuel into the cylinders, for the increased power generated to overcome the flywheel inertia of the mechanical drive train, and manifest itself as an increase in vehicle acceleration.

With the Earth's climate system, time lags are much more varied in input, and can have a wide range of duration. Sometimes, these time lags overlap and reinforce one another; sometimes they overlap and cancel each other out. When they all pile up in the same direction, be prepared for major events.

In signals and systems analysis there are basic input functions that can be imposed on any physical system. The output response to these inputs are a function of the systems mass, springiness and friction. The electrical analog to physical parameters is capacitance, inductance and resistance. Thus, it is possible to take any system, be it mechanical, thermal, hydraulic, or whatever, and construct an electrical analog or circuit, expose it to various inputs and observe what the resulting output would be. There are four basic input functions that can drive any system; a spike, a step, a ramp and a sinusoid. There is also the exponential input function which has no practical purpose other than to blow up systems.

A spike input function in systems analysis is a theoretical impulse of known amplitude, but with virtually no time duration.

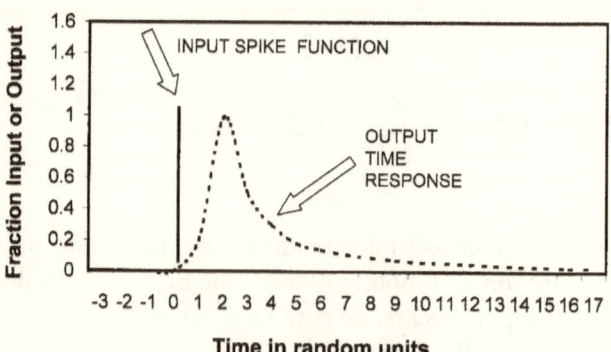

Fig 18a: Theoretical Spike Input and Output Response

In practice, the time duration is extremely short, but not zero. In geological terms it is instantaneous. In generational terms its input was about a few weeks. Its output or effect lasted only a couple of years. An example of this can be seen in the graph below. Within a 50 year period the stratospheric temperature had three spikes.

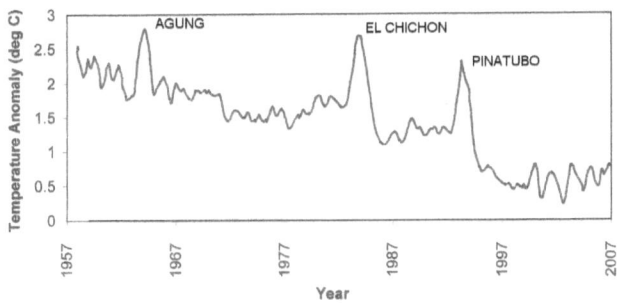

Fig 18b: Stratospheric temperature spike from volcanic eruptions

The stratospheric temperature spiked in 1967, 1982 and 1992 from eruptions of Agung, El Chichon, and Mt. Pinatubo, respectively.

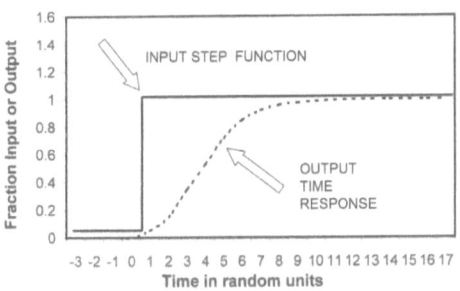

Fig 18c: Step Function Input and Ramp Function Output

Time lags or event delays should not be a surprise in nature. The hottest hour of a summer day is usually about 3PM, 3 hours after the Sun's zenith in the sky. The hottest days of the year are in July, about 4 weeks after the Summer solstice of June 21st. The coldest days of the year in the Northern Hemisphere are in January, about 4 weeks after the Winter solstice of December 21st.

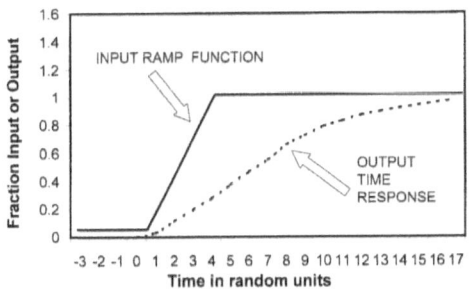

Fig 18d: Ramp Function Input and System Function Output

169

An example of a ramp function input and ramp output is the Northern Sea Ice Extent from 1900 to 2000.

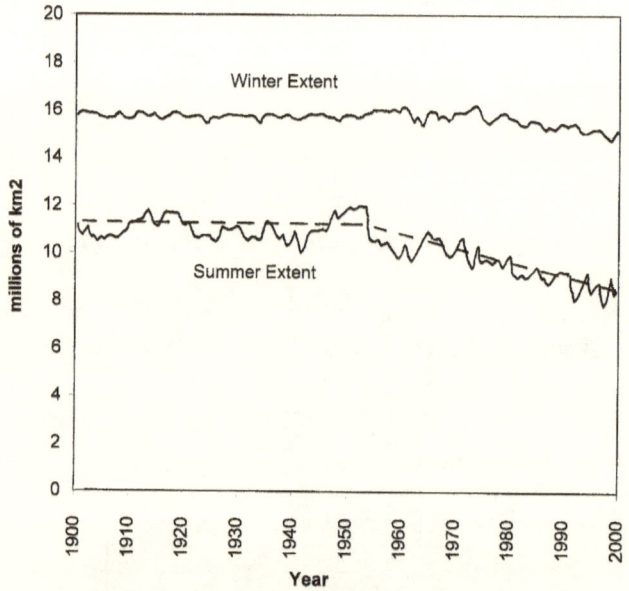

Fig 18e: Negative Step Function Input and
Northern Sea Ice Extent System Function Output
<http://www.arctic.atmos.uiuc.edu>

It is obvious the summer sea ice extent was relatively constant from 1900 to about 1953. Then the sea ice extent began a ramp down. A ramp output is typically the result of a step function input. That is: something wasn't happening, then suddenly and continuously it was happening; like opening up the drain valve on a swimming pool.

So far, we have discovered a plethora of cycles each individually traceable back to the Sun, Earth, Moon, planets, oceans, atmosphere, flora and fauna abundance. In this chapter we build upon these seemingly unconnected cycles and begin to see a pattern. Many of these otherwise disconnected cycles are actually coupled together. Take the figure below for example.

There are 12 plots of data. The top 3 originate from the Sun. The next 4 involve changes to the atmosphere. The bottom 5 show changes to the Earths' rotation and pressure and temperature. This arrangement shows signs of the Law Of The Triangle, again. The top 3 are like the input or cathode of a vacuum tube. The next 4 are variables being throttled like the grid in a vacuum tube. The bottom five are the outputs corresponding to the plate in a vacuum tube.

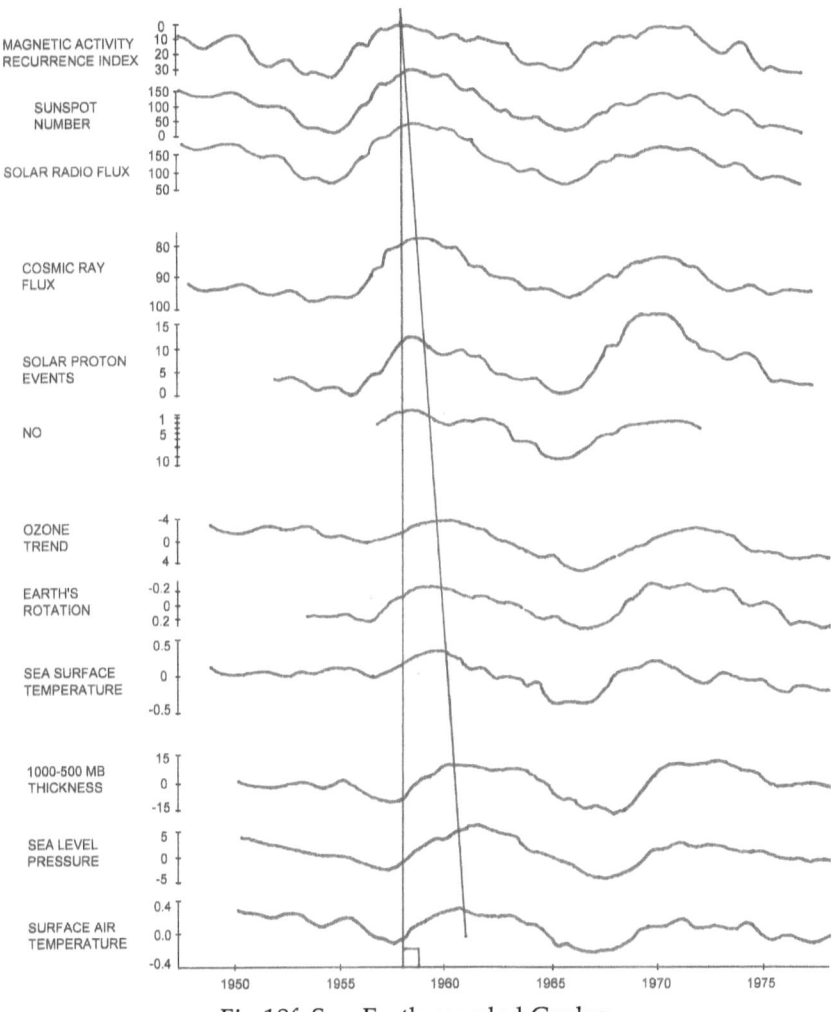

Fig 18f: Sun Earth coupled Cycles
(Climate, Fig 13-5)

There are two very important relationships of these 12 plots. The first and most notable is that they all seem to be related. The second is that there is a discreet time lag of the variables and outputs compared to the source input. This time lag varies from 1.2 years for NO changes, 1.5 years for Ozone changes, 1.8 years for sea surface temperature, and finally 2.4 years for surface air temperature. Please note that the Mauna

171

Loa CO2 concentration ramp curve is not shown, because it is simply not relevant.

The solar wind speed has also been found to lead cosmic ray flux variation by about two years. Solar geomagnetic activity leads Earth global temperatures by 4 to 8 years.

The 44.77 Year Sun Tide Cycle

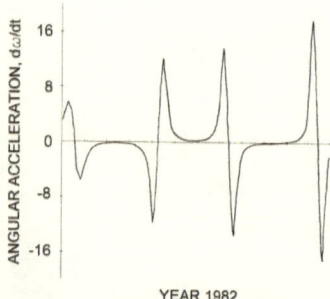

Fig 18h: Angular Acceleration of Tidal Planets
(Landscheidt T, Sun Earth Man, Urania, 1989)

It is the planets Jupiter, Venus and the Earth that control the Sun-tide period or sunspot cycle on the Sun. (Bollenger, Clyde J., A 44.77 Year Jupiter-Venus-Earth Configuration Sun-tide Period in Solar Climatic Cycles, Academy of Science For 1952). This is why J-V-E are called the Tidal Planets. They control the sunspot cycle. It is hard to believe that our tiny insignificant little Earth can have a contribution to the climate of the Sun. Should it be any wonder then, that the Sun dictates our climate as well?

The 1470 Year Cycle

The 1470-year cycle has been documented in various different ways. It is referred to as the Ice Rafting Debris (IRD) Cycle, Dansgaard-Oeschger Oscillation warming events, Heinrich ice-discharge or melt-water events or cooling episodes, and Younger-Dryas events. Roughly 1,400-year to 1,500-year periods include the Great Lakes water level cycles, North American Pollen Database cycles, Mayewski solar waxing-waning cycles. All these cycles have been documented by the changing oxygen-18 isotope concentrations in sea-floor and land sediments, and ice core samples.

The 1,470-year cycle just happens to represent the Least Common Denominator of the 87-year Gleissberg Cycle and 210-year deVries Cycle. This has led to the possible solar origin of the 1,470-year glacial climate cycle. (Braun et al, Nature Letters, **438**, 208-211, 10 November 2005). Two results come out of this coupled model study. The first, shows the Gleissberg cycle amplitude modulated by the 11-year Schwabe cycle results in peaks at 210, 86.5, 43.2, 12.4, 10.8, and 9.6 years. The second, shows the Gleissberg Cycle modulated by the DeVries Cycle resulting in peaks at 210, 147, 105, 86.5, 61.3, 43.2 and 28.8 years.

The 934-Year Jupiter-Saturn Resonance Cycle

There is a distinct 934 year climate cycle. It is driven by the orbital resonance of Jupiter and Saturn. This is elaborated by P. A. Semi's paper on Orbital Resonance and Solar Cycles <https//arxiv.org/ftp /arxiv/papers/0903/0903.5009.pdf>

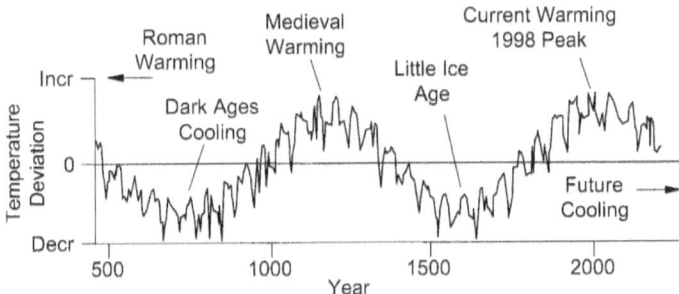

Fig 18g: Temperature Deviation For The Last 934 Years
<www.earth-climate.com>

As can be seen above, the current warming cycle has peaked out and is heading back down again.

Conclusion

Should it be any surprise that the two largest planets, acting by Keplerian Mechanics tugging away at the Sun, should have such a significant impact on our Earth?

173

"It is always more difficult to fight against faith than against knowledge"

- Adolf Hitler

Chapter 19: Weather Forecasting

Prepare yourself.

Every day there is someone on Wall Street predicting the next crash being just around the corner. When the crash does happen, that lucky someone will get accolades for being the genius with the foresight to have correctly foreseen this event of the future. Actually, it wasn't genius at all. It was simply the luck of the draw. The inevitable event just happened to occur on his or her watch.

People have been predicting the end of the polar ice caps for decades ever since the 1960s. One day it will happen, or maybe not. In the 1970's the media and some scientists were predicting the next Ice Age was upon us. In the 1990's Al Gore and his minions were screaming about global warming and our planet being turned into a cinder. By the end of the 2000's, it looked like temperatures had stopped climbing, the Sun was going to sleep, and cooling might be resuming. All this stuff has happened before, and it will happen again, with or without mankind's involvement.

Finally, consider the heretofore unthinkable: In the not too distant future CO_2 levels will reverse direction and head back down. This, too, is a cycle, which the global warming cultists, in general, are either totally unaware, or worse, they know, but are not willing to share.

The Meteorological Office ("The Met")

The Met Office has recently occupied its new, spacious facility at Fitzroy Road, Exeter, Devon. Apparently, the facility is so big, it does not warrant a house number for the street address, just like NASA (no street to be used, because it is so important).

Fig 19a: The UK Met Office, Exeter, Devon
<http://en.loadtr.com/Met_Office-477025.htm>

'The Met' in the United Kingdom started out in 1854 to serve the weather forecasting needs of mariners. After the invention of the airplane, it grew to include serving aviators and became an arm of the Ministry of Defense. Over the years the purpose of the Met Office has mutated from a data gathering organization serving mariners and aviators into a long range weather forecasting organization run by climate zealots.

The Met Office has a grotesquely huge annual budget of £170,000,000, and a staff of 1,500. It uses a £30,000 supercomputer that has the carbon footprint of a 1,000 homes. And yet with all this money, manpower and resources, The Met consistently manages to get its forecasts wrong. Why? The answer is simple when you realize The Met Office has hitched its prognostications onto the miniscule effects of a trace atmospheric gas rather than the colossal celestial drivers of real climate change.

This change in purpose started between 1983 and 1991 when John Houghton was the Director General. It was he who attended the first World Conference on the Changing Atmosphere and later became the first scientific chairman of the IPCC. It was Houghton who brought into existence a new Met Office unit called the Hadley Centre for Climate Prediction and Research, which now has its own staff of 200.

The Met Office is now headed by Robert Napier, who was previously head of the UK branch of the World Wildlife Foundation from 1999 to 2007. Napier has many ties to green organizations, conservation groups, and willing to work with any organization that believes man is ultimately to blame for climate change.

Weather Action

While the Met Office regularly gets it's long range forecasts wrong, Piers Corbyn at Weather Action gets his long range forecasts right about 85% of the time. Armed only with a laptop computer, access to public data and a degree in Astrophysics, Piers Corbyn consistently out-forecasts the Met Office.

Weather Action is located in a modest office at Delta House, 175-177 Borough High Street, London, a few blocks from London Bridge. Delta House is a dated, 4-story office building with many tenants, the largest space being 1,600 square feet. I do not know, if Weather Action is at the front of the building with a window, or has a more modest interior cabin.

Fig 19b: Weather Action at Delta House, London
< https://www.bca.uk.com/centre/lenta-business-centres-17>

It is telling how the size and quality of the accommodations of The Met Office and Weather Action are inversely proportional to the quality of the forecasting product they generate. It is also ironic to compare the amount of glass in both pictures above. You would think with all the glass windows at the Met Office they would know what the weather is outside.

And finally this from WeatherAction's web site:

"The unique power of the forecasts has also been proven by the profits on Scientific Weather Bets with William Hill at odds and verification organized independently by the UK Met Office.

Bets and notional bets can be used to estimate Forecasting Power which is the % profit (or negative for losses) on stakes that would come from bets placed at fair odds. For general long range forecasts for the three most extreme recent seasons, namely Summer 2007, Summer 2008 and Winter 2008-09 The Met Office long range forecast Power is minus 100% (ie Met Office long range forecasts failed in all three cases) and WeatherAction (Solar Weather Technique) scores about plus 500%.

In 4,000 Weather Test Bets over 12 years with William Hill, Weather Action forecasts made a profit of some 40% (£20,000). The Odds were statistically fair and set by the Met Office before being shortened by William Hill by a standard 20%; the results were then provided by the Met Office for William Hill to settle each bet.

Piers Corbyn was excluded by the bookies from such account betting in 2000."

Translation of the above: Piers Corbin of Weather Action is so accurate, so often, with his forecasts, he is no longer allowed to wager in the Scientific Weather Bets lottery. The Met, however, with is poor track record is apparently not held to this restriction. The lottery is more than happy to take the Met's (ie: taxpayer's) money.

Long Range Weather Forecasting

Depending on what you read, weather forecasting has three ranges, including a 5-day short range, 6 to 10 day medium range, and long range greater than 10 days. Not surprisingly, even these multi-day forecasts often turn out to be wrong.

Long Range Weather Forecasting (LRWF) has become a new industry, which caters to commodity traders, power companies, farmers, transportation businesses, municipal road departments, or anyone else who needs to know far in advance what the weather is going to be, not next day or next week, but next month and next year. This new LRWF technology is not based on the usual methodologies utilized by the meteorological establishment.

Harris Mann Climatology

LongRangeWeather Harris-Mann Climatology is another LRWF run by climatologist Cliff Harris and meteorologist Randy Mann. On the home page of their website they provide a detailed temperature account from 500BC to the present. A simplified graph of that account follows.

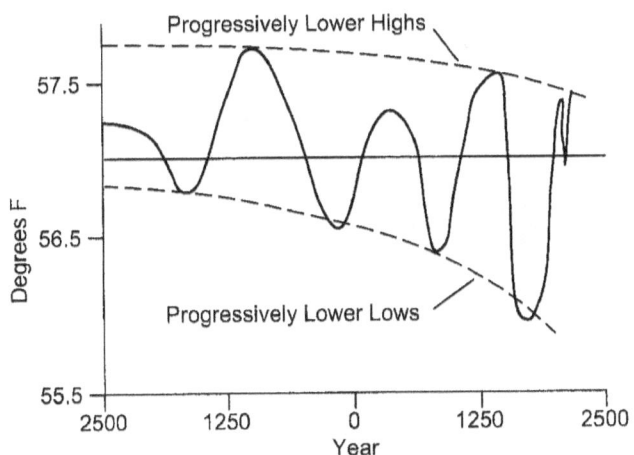

Fig 19c: Lower Lows and Lower Highs for last 5,000 years
(Based on Harris-Mann Climatology)

The interesting tell of this curve is the progressively lower highs and lower lows that have occurred over the last 4,500 years. This corresponds closely with R. B. Alley's Ice Core Curve in Fig. 21c.

HMC points out that we are in the present 102-year cycle which started out in the Dust Bowl year of 1936 and is due to complete by 2038. They also predict "that we could see a huge stock market crash, global depression, new outbreaks of unknown disease, and perhaps, a third World War if we stay warm and precipitation levels fall on a global scale sometime between 2020 to 2038."

The Long View

Summarizing from his website, www.thelongview.com.au: "Kevin Long has been a mechanical engineer since 1977, and is Owner/Manager of K.E.V. Engineering, based in Bendigo, Australia. Kevin's interest in weather systems, climatology, agriculture and catchment management stems from his roots on the land. Kevin has been a keen observer of meteorological data and weather forecasting systems throughout his life. He has published quarterly seasonal weather predictions for Central Victoria since 2004, based on his study of sea surface temperature patterns and other climate indicators. These forecasts have been consistently close to the mark, without the "positive spin" so often put by government agencies and commercial businesses."

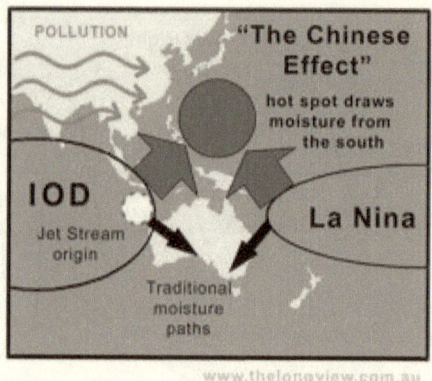

Fig 19d: The Chinese Effect
<http://thelongview.com.au/chinese-effect.htm>

Again summarizing from Kevin's website: "In 2006, Kevin Long was the first to name and identify the impacts of "The Chinese Effect" on Australia's weather. Kevin believes this to be the major tipping force behind climate change for south-eastern Australia. "The Chinese Effect" is the annual warming of the seas north of New Guinea due to Asian aerosol pollution. The hotspot draws moisture in from south of the equator, thus reducing the moisture available to the Australian rainfall systems. The hotspot has been very apparent since 2004. This significant and increasing effect on Australian weather has not been recognized by

180

governments and water management authorities. We are not dealing with a conventional "drought" that is going to "break". We are witnessing distinct "climate change" due to a new pattern of equatorial moisture flows.

Kevin's current long range forecast covers the period from 2010 and 2028 <http://thelongview.com.au/documents/NEW-CLIMATE-CYCLE-2010-TO-2028-Kevin-Long-v1.pdf>. That is impressive; an 18 year forecast!

The Old Farmers Almanac

No long range weather forecasting medium is probably more well known than the Old Farmers Almanac. The Almanac's weather predictions are supposedly based on solar activity, astronomy cycles and weather patterns. The research of its founder, Robert B. Thomas culminated with developing a secret forecasting formula first created in 1792, which is after some recent tweaking is still used today. Like the formula for Coca-Cola, few people have seen the forecasting formula.

In 2008, the Almanac stated that the earth had entered a global cooling period that would probably last decades. The journal based its prediction on sunspot cycles. Contributing meteorologist Joseph D'Aleo said that "studying these and other factor suggests that cold, not warm, climate may be our future."

Chicago's WGN9 noted in 2019, according to a University of Illinoios study, "The Old Farmers Almanac claims a forecast accuracy of 80 to 85 percent, but in reality its forecast accuracy here in the Midwest is on the order of 50 to 52 percent."

Radio Weather

In 1901 radio transmissions began. In 1923 the Bell System started a program to measure disturbances in radio transmissions. By September 1927 interference of radio transmissions were found to be coincident with large sun spots and auroras. During WWII, what was first thought to be the Nazis jamming radio signals, it was found that there was a correlation between the radio interference and geomagnetic and solar activity.

In 1940 John H. Nelson, a radio technician, concluded with the great magnetic storm that began on March 23, 1940, that a special alignment of Venus, Earth, Saturn and Pluto were involved. On April 19, 1946, Nelson was appointed Solar Researcher for RCA Communications at their NYC Headquarters. During his tenure he met Jack Clark a radio frequency engineer for Press Wireless and learned that the three most distant planets – Uranus, Neptune and Pluto were involved with the development of reliable magnetic storm predictions.

Nelson concluded that at least two of the fast moving planets Mercury, Venus, Earth and Mars and one or more slow moving planets Jupiter, Saturn, Uranus, Neptune and Pluto must be involved. During 1967 he made 1460 forecasts with an accuracy of 93.2% (Dewey page 182). One can only conclude that the magnetic field of the Sun interacting with the magnetic fields of the planets as they orbit about the Sun can modulate and interact with the magnetic field of the Earth to cause our radio interference.

Inigo Jones

The theory of Inigo Jones, an Australian weather prognosticator, is based on the planets Jupiter and Saturn shielding the Sun from the magnetic field thru which the Solar System moves. His forecasts involved predicting droughts which usually coincided with J-S shielding a spotless Sun. Though Jones spent his entire forecasting career at the Crohamhurst Observatory and had a rather successful track record, his predictions commonly paralleled those which were based on lunar cycles.

Conclusion

The tools available to today's meteorological forecasters are very limited. They are only good for making forecasts of only a few days into the future. Forecasts attempted beyond this point become impacted by Lorenz' Chaos Theory (ie: The further into the future you look, the less accurate the forecast.) When extending weeks, months and beyond classical forecasting methods disintegrate.

It is time for improved forecasting methods that can reach far into the future and with accuracy. And that time has arrived.

"When everyone is thinking alike, no one is thinking at all".
— General George S. Patton

Chapter 20: Climate Forecasting

After seeing how weather forecasters cannot even predict the next day's weather with any accuracy, you wonder what would be the sense of them trying to predict the weather years or decades in advance using the same methodologies. Well the answer lies in the use of prognostication tools that are totally different; tools, such as, analyzing sun spots and solar storms, rather than observing the movements of low/high pressure areas and measuring CO_2 concentrations.

Sloan & Wolfendale

Scientific American published an article on 11-11-13, Cosmic Rays Not Causing Climate Change, with the subheading Cosmic rays have played at most a very small part in global warming, new research finds. The article goes on to say that the record of global temperatures studied goes back to 1955. This study examined Solar Cycle 22 and concluded there is little relationship between Cosmic Ray Flux (CRF) on the Climate.

Shouldn't this article and study stand out immediately as lacking in a number of respects? The first questions I have are: why didn't these two scientists look at and address the data of solar cycles 1 thru 21, and only look at cycle 22? What are they trying to hide? The second question I have is: Is there any data going back further than the modern twenty-something solar cycles? The answer to that question is: Yes. The third question I have is: Does anyone really have any respect for this woefully inadequate piece of research?

I did not want to give away the ending to the story on CRF started in the Chapter on Earthly Influences.

With less solar wind, more galactic cosmic rays can penetrate the magnetosphere and pass thru the atmosphere. This action can be compared to that of the grid in a vacuum tube acting as a valve and permitting more plate current to flow with less grid voltage. (Just thought if there were any old vacuum tube enthusiasts out there, they would get a kick out of this analogy.)

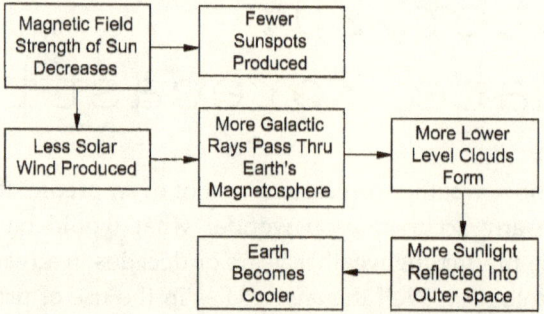

Fig 20a: The Solar Climate Relationship

The figure above says it all. Sunspots are just a manifestation and signature of magnetic field strength; the fewer spots, the weaker the sun's magnetic field strength. Weaker magnetic field strength means less solar wind, resulting in more galactic cosmic rays passing thru the Earth's Magnetosphere, which forms more low level clouds, allowing more sunlight to be reflected back to outer space, causing a cooler Earth. CRF is global cooling.

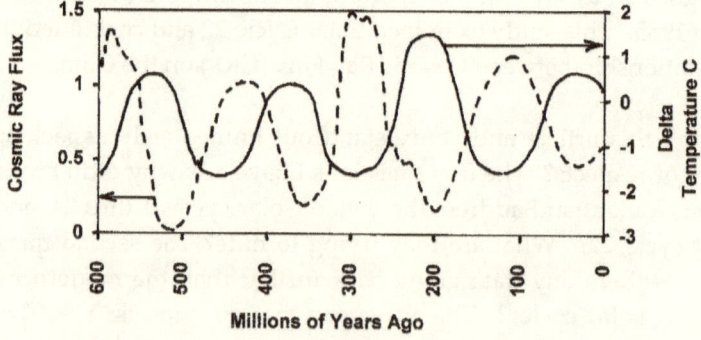

Fig 20b: CRF vs Temperature for last 600 million years

It has been contended in past IPCC Reports that the effect of CO2 on the atmosphere is about 1.5 w/m2. A recent study ?? shows that this effect my be a tenth of past predictions, or 0.15 W/m^2. Changing cloud formations could result in a median change of about 24 w/m2. Therefore, the worshipers in the church of AGW will be saddened to hear that the effect of changing cosmic ray flux entering the atmosphere is about 24/0.15 = 180 times more significant than the effect of CO2 on a w/m^2 basis. Let me repeat that: the effect of changing cosmic ray flux entering the atmosphere is about 180 times more significant than the effect of CO2.

Duhau and de Jager

Solar variability is dominated by a buoyant rising toroidal and a surface poloidal magnetic field component occurring in a layer of about 30,000-km thickness situated 200,000-km below the solar surface. This is the solar dynamo. This is the engine that determines how warm and how cold the Solar System has been, is, and will be.

The solar dynamo operates in three Grand cycles. The Grand Minimum which last occurred between 1620 and 1724 (The Little Ice Age), the Regular Oscillations (1724-1924), and Grand Maximum (1924-2009) (Duhau, S. and de Jager, C., The Forthcoming Grand Minimum of Solar Activity, Journal of Cosmology, Vol. 8, 1983-1999, 2010). That's right folks, the Grand Maximum ended in 2009. Care to guess what's next? Yes, that's right - the next Grand Solar Minimum has started, and with it the cooling that comes along with it.

Solar activity is presently going through a transition period (2009-2013). This will be followed by at least one very low Schwabe Cycle, which now seems to be unfolding before our very eyes. This is the harbinger of the forthcoming Grand Minimum, which will most likely be of the long type. Translation: Let your children know they should be prepared for colder winters for decades to come. Also, be warned that no amount of miniscule CO2 warming increases are going to slow down this massive cooling.

The starting and finishing points of these Grand Cycles appear to be governed by the upper bands of the Gleissberg and deVries Cycles. A graphical representation of these two Cycles is shown in the figure below.

185

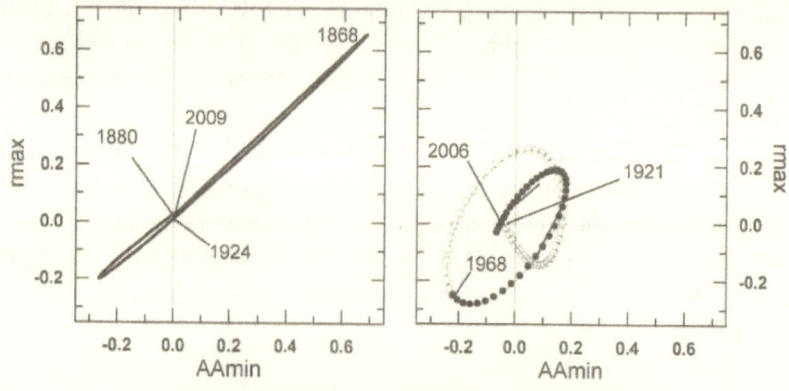

Fig 20c: Gleissberg Cycle Phase Diagram
(Duhau and de Jager, 2010)

You do not have to be concerned with the "AAmin", and "rmax" above, or the pattern plots. Basically, what these diagrams say are: the Next Little Ice Age has started.

Let us hope this is not accompanied with increased aerosol cooling of the atmosphere from recent increases in volcanic activity, and a prolonged negative PDO. Otherwise, cooling of a much more serious nature could be on the horizon.

<u>Adriano Mazzerello</u>

Unfortunately for us, it appears a prolonged negative PDO is on its way according to Adriano Mazzerello (Mazzerello, Adriano, Solar Forcing of Changes in Atmospheric Circulation, Earth's Rotation and Climate, The Open Atmospheric Science Journal, 2008, 2, 181-184).

This study examines 150 years of historical data and creates a new model different from the usual Global Climate Models (which use only greenhouse gases as a climate driver). Mazzarello's model incorporates a Sun-atmosphere-Earth model that shows an increase in 'solar corpuscular activity', causing a deceleration of the Earth's zonal atmospheric circulation, causing a deceleration of the Earth's rotation, which in turn causes a decrease in sea surface temperature (SST).

Mazzrello's study states a 60-year cycle has been identified, which suggests a SST cooling that began in 2005. Based on this study, you can expect a bottoming out of temperatures around 2035.

Global Weather Oscillations

Global Weather Oscillations (GWO) is yet another LRWF, formed as stated in their website "with the specific understanding that almost all climate and weather events occur in cycles, and it is the Primary Forcing Mechanism (PFM) that controls the Earth's Climate Pulse and climate-weather oscillations. GWO utilizes this knowledge to forecast cycles of weather/climate up to 4 years in advance, and climate change cycles well beyond 100 years."

One of GWO's more memorable predictions was in 2011, where it predicted a "hybrid" hurricane, one of the strongest in 50 to 100 years, hitting the northeast USA in the last quarter of 2012. We all know that prediction became Superstorm Sandy, which was a perfect storm, the hybrid between a Category 1 hurricane and a 'NorEaster. Good call GWO. And where was AccuWeather or The Met on this?

Qian and Lu

Qian and Lu have studied solar radiation and its effect on Pacific SST and resulting Global Mean Temperature (GMT). (Qian W H, Lu B, Periodic oscillations in millennial global mean temperature and their causes, Chinese Sci Bull, 2010, 55: 4052-4057, doi: 10.1007). Their findings show that GMT has four cycles: a 21.1 year cycle that peaked in 2002, a 62.5 year cycle that peaked in 1998, a 116 year cycle that peaked in 1994, and a 194.6 year cycle that peaked in 1998. These four cycles are forced by and lag behind solar radiation variability by 1-2 years. They also determined the 65-year oscillation lags behind the PDO Oscillation.

The 21.1 year peak has an amplitude of 0.016C anomaly, the 62.5 year peak has an amplitude of 0.047C anomaly, the 116 year peak has an amplitude of 0.052C anomaly, and the 194.6 year peak has an amplitude of 0.069C anomaly. When these four peaks reinforce in the year 1998 their total amplitude is 0.184C. However, when the trend-removed original temperature series is used, the amplitude becomes 0.3C. Thus, the effect of solar radiation on the Pacific SST alone accounts for about half of the 0.6C temperature increase the IPCC is blaming on man for the last 150 years.

The formula for the 21.1 year cycle is $-0.016\cos(0.3t+5.2)$, where "t" is the year AD. If you plot this yourself, you will first have to divide the year by Pi (3.14159). The formula for the 62.5 year cycle is +0.047cos(0.1t+0.22). The formula for the 116 year cycle is +0.052cos(0.05t-1.13). The formula for the 194.6 year cycle is +0.069cos(0.032t-1.7). The plots of these four cycles are below.

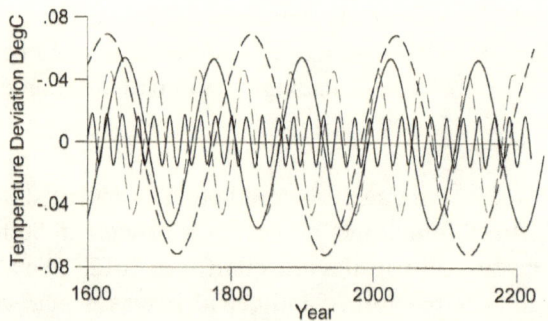

Fig 20d: The Four Individual Qian and Lu Cycles

What is most revealing is a plot of the sum of the four individual cycles together, which are shown in the composite cycle below. Suddenly, when these four individual cycles are combined, they take on a whole new, revealing trend.

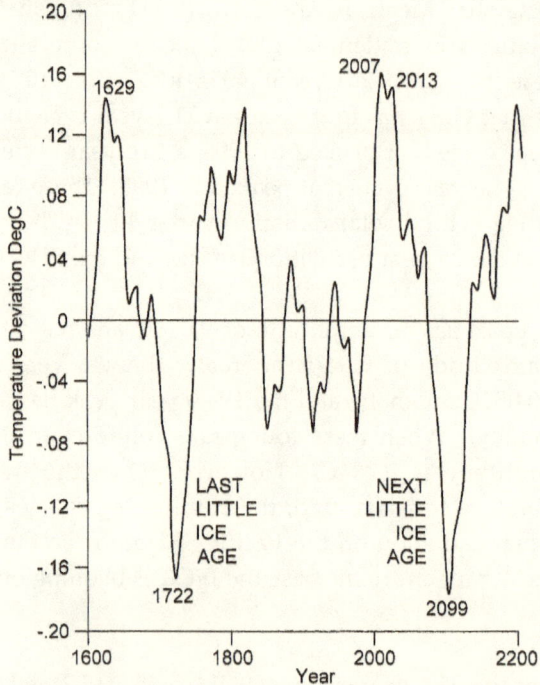

Fig 20e: The Composite Four Qian and Lu Cycles

188

First, note above how temperatures bottom out in 1722 and again in 2099. This appears to be a recurring 377 year pattern. Also, note the 2007 and 2013 double peaks, which occur again about 381 years after the 1629 peak.

The four Qian and Lu cycles almost peak out in 1998, the year that NASA GISS proclaimed as the hottest year ever recorded back then. Do you think NASA GISS credited the natural forcing of solar radiation on Pacific SST for the 1998 record? And please notice the correspondence of the Q-L composite spanning hundreds of years with the Vostok Ice Core Temperatures spanning hundreds of thousands of years; multiple ever-decreasing temperatures followed by a sudden violent increase.

Koelle Future Temperature Extrapolations

Dr. Dietrich E. Koelle has written a paper dated 25 January 2015, entitled Current Warm Period is No Anthropogenic Product. In this paper he analyzes three warming cycles: The 1,000 year Suess, 230 year de Vries, and 60 year Ocean.

Like the Ozone Trend Panel cherry picking ozone trends as depicted in Figure 5e, the Global Warming Crowd also likes to cherry pick temperature trends. The GWC took the 1970s as a starting point and projected a linear increase as far as the eye can see as depicted in the figure below. The problem is: the temperature trend is the 60 year ocean oscillation cycle. As borne out by satellite measurements, the warming stopped around 1998, which not surprisingly corresponds with the peak in the 60 year ocean cycle.

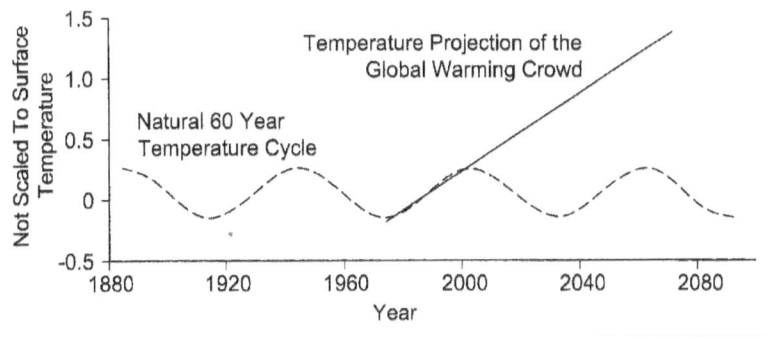

Fig 20f: Global Temperature 1880-2080
(based on Koelle1.jpg)

Then Dr. Koelle next looked at the 60 year and 230 year cycles from 1860 to 2020. Notice how the 230 year cycle bottoms out around 1900 and peaks at about 2000.

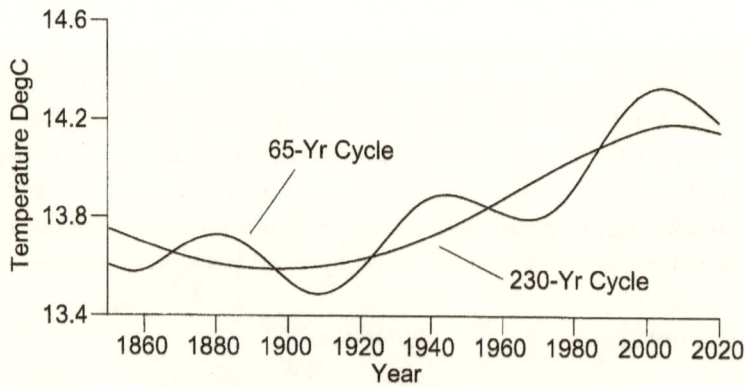

Fig 20g: Global Temperature 1860-2020
(based on Koelle2.jpg)

Then Dr. Koelle reviews the last 3,200 years of the Suess and de Vries cycles. Notice how the 1,000 year Suess cycle coincides with the temperature peaks of the Medieval, Roman and Minoan Warming Periods.

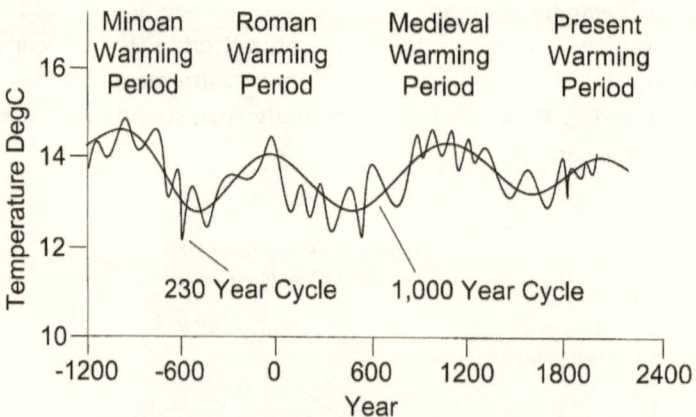

Fig 20h: Global Temperature -1200-2200
(based on Koelle3.jpg)

Finally, Dr. Koelle extrapolates the 1,000 year Suess and 230 year de Vries cycles out to the year 2600. Notice the 1.0C drop in temperature from today to the year 2600.

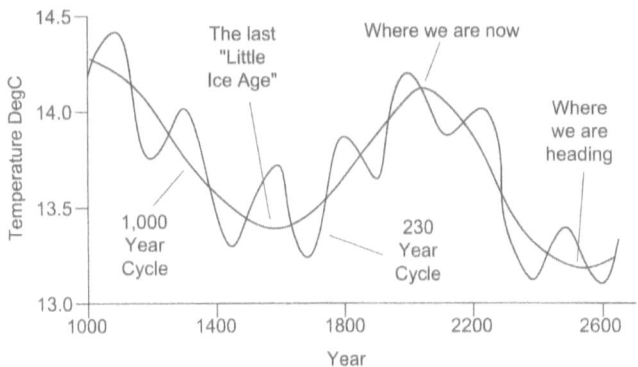

Fig 20i: Global Temperature 1000-2600
(based on Koelle4.jpg)

H. I. Abdussamotov

Habibullo Addussamotov, Dr. Sc. is the Head of Space Research Laboratory of the Pulkovo Observatory in St Petersburg, Russia. Dr. Abdussamatov has contributed Chapter 17, The New Little Ice Age Has Started, in the book Evidence-Based Climate Science (Second Edition) (doi.org/10.1016/B978-0-12-804588-6.00017-3)

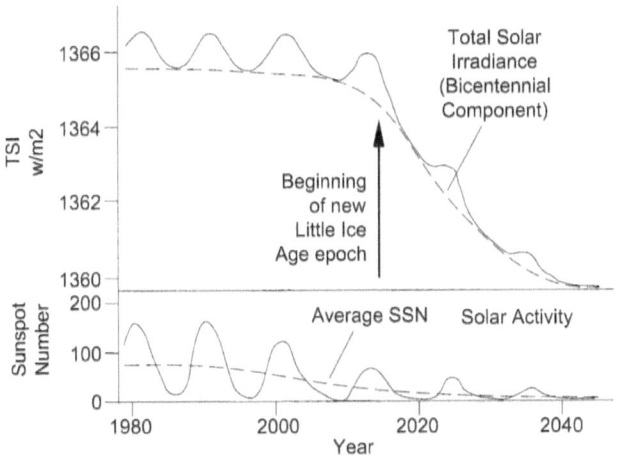

Fig 20j: Total Solar Irradiance 1980 to 2050
(Based on Climate Science, Chapter 17)

As can be seen above the total solar irradiance (TSI) has been relatively flat starting from 1980 to about 2000. Thereafter, the TSI begins a steep decline ending about 2050. During this plunge the TSI is projected to decrease about 6 watts/m2. This correlates with a global land-sea

191

temperature decrease of about 3C or 5C, or up to 9F! Should there be a weakening of the Gulf Stream, western Europe and the eastern parts of the United States and Canada will experience even further drops in temperature.

391-year Supersecular Cycle

Between the period of 1100AD to 2100 AD, there is a varying wave of secular energy from the Sun. Further, the climate of the Northern Hemisphere has changed in step with changes in the Sun's long term variations. Extending this cycle reveals our climate will experience a temperature minimum around 2030 and a maximum around 2130. (Climate, page 433)

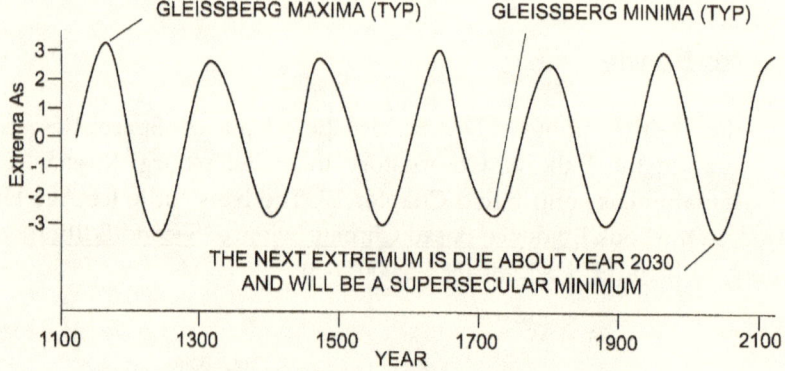

Fig 20k: Solar Impulses of the Torque
(based on Climate, Fig 25-10)

The above figure is almost a perfect sine wave. However, considering a thousand years of data, and 12 rather consistent periods, it is safe to predict future 13th trend with a certain degree of probability. The quantification of this confidence is represented by what is called a Pearson Chi-Squared Test, which calculates the probability of an outcome based on a frequency distribution of certain events. In this case the certainty or "p" of an event occurring or not occurring like the flip of a coin (a 50% chance) happening 12 times in succession is equal to $(1/2)^{12}$ or 0.000244. Or said another way, the odds are 1/0.00024, or 4,167 to 1 that the next cycle will be about the same.

The 9-year, 72-year, and 230-year Arctic Pulses

David Dilley is the CEO of Global Weather Oscillations (GWO) and author of the book Global Warming – Global Cooling, Natural Cause

Found. I strongly suggest you download and read every word of this enlightening, scientific work, which is available at no charge. It is a 23 page document which contains the entire explanation of where the climate has been, and where it is heading.

The Earth's climate rhythm to the strongest degree is controlled by the motion of the Sun, Moon and Earth. These interactions are sub-cycles to the much longer Milankovich Cycles. Mr. Dilley calls this interaction a Lunisolar Precession. It has three distinct cycles of 9-years, 72-years and 230-years. The 9-year cycle is a harmonic of the 18-year Lunar Nodal Cycle.

These cycles create warm water pulses into the Arctic Ocean from the Atlantic. The strongest pulses are a 72-year cycle. The most recent pulses occurred in 1990, 1998, 2007. These can be seen in the following graph.

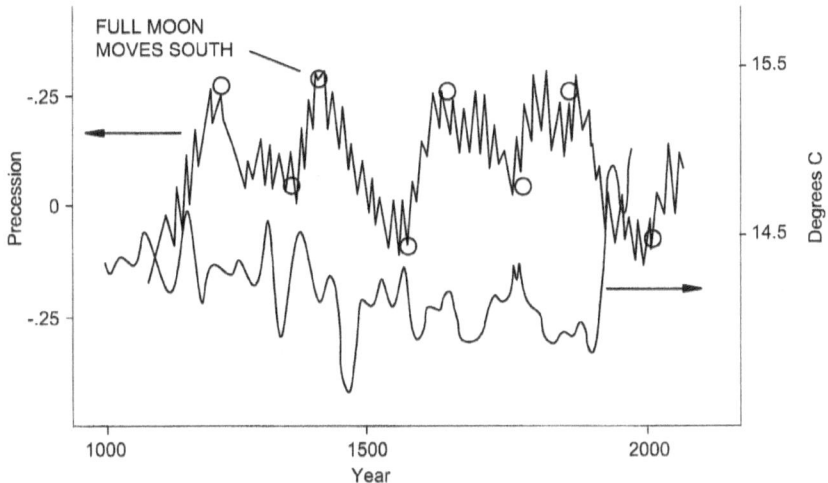

Fig 201: Lunisolar Precession 959 to 2070
(Natural Climate Pulse, D. Dilley, 2012)

From the graph above, you can see the 230 year cycle of the full moon with its four peaks and four valleys. In the current cycle there were warm periods from 1930 to 1940 and 1998 to 2008. In 2008, the Lunisolar Precession began its ascent signaling the beginning of what Mr. Dilley calls the Phase I Global Cooling. Phase II Global Cooling is estimated to begin around 2020, with the coldest temperatures occurring between 2023 and 2050.

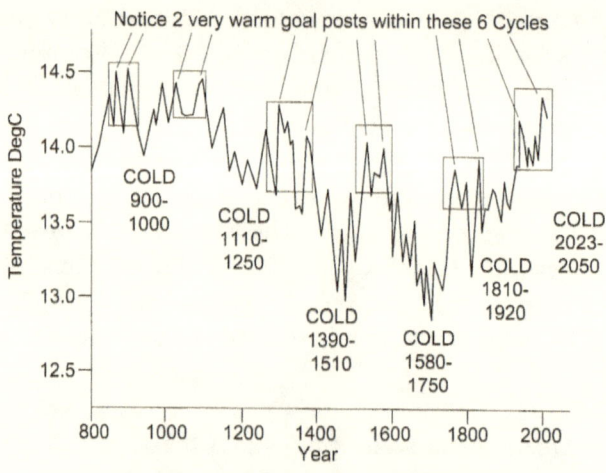

Fig 20m: 230 Year Cycles 800-2000
(Natural Climate Pulse, D. Dilley, 2012)

In the 230 year cycles graph above, there are six distinctive warm and cold cycles. There is a 1.7-degC or 3.06-degF difference between the 14.5C maximum and 12.8C minimum. You can see the 1,000-year Suess cycle starting out around year 1000 and ending in year 2000

Predictions

Normally, it is not a good idea to make predictions. Al Gore, Hollywood types and other global warming zealots are constantly making groundless predictions. These predictions which have no basis in fact whatsoever ultimately do not pan out. Predictions, such as, 'in 10 years all the fish in the oceans will die', 'the Earth will enter irreversible runaway greenhouse gas heating', 'the Arctic Ice Cap will disappear in 7 years', 'polar bears will become extinct', 'sea levels will rise flooding all coastal cities', etc.

However, after having presented all the evidence in this book, it is obvious in which direction the Earth's climate is heading. Understanding that it is the planets which regulate the Sun's climate, and the Sun's climate and the Moon which dictate the Earth's climate, there can be only one logical conclusion.

Based on satellite temperature measurements, in 1998 the Earth entered a cooling phase. This cooling phase looks like it will go on for decades. Each successive year more and more low temperature records will be set, documenting the progression into the next Grand Solar Minimum – the Landscheidt Minimum. More shockingly, CO_2 levels will begin to reverse their climb and descend to their customary cyclical historic low levels. This isn't mindless conjecture. This is acknowledging the cyclical nature of climate for decades, centuries, millennia, eras, and epochs. This is the understanding of the immutable, totally predictable laws of celestial mechanics.

Folks, temperature-wise, it is all down-hill from here! Anecdotes are not proof, but all the data is piling up in the same direction and homing-in on the same time frame. It is becoming clear with every subsequent season, that the climate is indeed changing; changing in the direction of cooling temperatures. Rather than worry about warming which has all but disappeared, it is time to prepare for cooling. Many more people perish from cold than they do from warming. With all the continued nonsense and concern for non-existent warming, the world is ill-prepared for this return to another Little Ice Age.

Interglacial Warmup Periods are about 11,000 years long. Ice Ages are about 90,000 years long. We are at the very end of the present IWP. After this comes centuries and millennia of cold, unforgiving, unrelentling, life killing, cold. Enjoy the warmth while you can.

Planetary and lunar cycles are sufficient to change the climate noticeably. Atmospheric and oceanic cycles also can change the climate significantly. Volcanic eruptions can suddenly inject aerosols into the atmosphere reflecting beneficial solar radiation. The real concern is that it is plausible that all these factors may within the next couple of decades happen all at the same time. If this confluence of events comes to pass, we as a civilization may be confronted with our most severe test of survival. And no pithy amount of temperature increase from supposed global warming from CO_2 increases will alter this inevitability.

Summary

Rather than summarize the points discussed above, I have listed the points by date, chronologically as follows.

1994: 116-year cycle of Global Mean Temperature (GMT) peaks

1998: 62.4–year GMT cycle peaks

1998: 194.6-year GMT cycle peaks

2002: 21.1-year GMT cycle peaks

2005: Negative PDO begins

2008: Lunisolar Precession begins ascent signaling Phase I of Global Cooling

2008 to 2024: Increased Earthquake activity

2009: Grand Solar Minimum begins

2020 to 2030: Eruption of Climate Changing Volcano Predicted

2024-2050: Coldest temperatures of this GSM

2035: Negative PDO bottoms out

2080: CO_2 levels return to their historic lows of about 250 ppm.

2100: The Composite four Qian and Lu cycles bottom out.

"Man's most valuable trait is a judicious sense of what not to believe"

– Euripides

Chapter 21: Global Warming Theory

Global warming theory is unique. In science, all it takes to disprove a theory is one experiment that fails to reproduce results or one fact that does not fit. When Albert Einstein postulated his Theory of Relativity, his fellow physicists recoiled in horror. In response, one hundred of his peers penned a letter denouncing Einstein's views. When presented with this letter, Einstein said "why are 100 needed? Isn't one enough?"

In global warming theory, facts and experiments proving the theory wrong continue to accumulate, unabated. Yet, regardless of how many facts and experimental results show that the theory is flawed, the theory continues to survive unscathed. It's like the Terminator, or Yul Brenner in Westworld. No matter how many times it is killed off it returns to terrorize again and again. How can this be?

Global warming theory is driven by the MSM starved for bad news to keep ratings up and ad revenue coming in, politicians needing a crisis with which to lead us to safety to curry votes, a cabal of scientists eager to burn up billions of dollars of readily available public funds to busy themselves while keeping dissenting scientists at bay, corporations eager for additional revenue from the new alternative energy cottage industry, sustainability and the green movement that owe their very existence to this theory, and an environmental movement that has metastasized from a group of truly concerned individuals deeply committed to protecting nature to a collection of misfits, anti-capitalists, anti-industrialists, Socialists, Communists, malcontents, psychotic tree huggers, and elitists who are antagonistic of anything or anyone that has to do with profit, comfort, freedom or security.

Global warming theory isn't science. Global warming theory is influence, money, power and control. Global warming theory is a religion requiring complete, unquestioning faith. Otherwise, you could be accused of being a 'denier'.

Global warming theory concentrates singularly on human influences, with only incidental acknowledgement of any natural influences, and for good reason. If the comparison was made between human and natural influences, the titanic natural forces in total control of our climate would be revealed and the global warming introduced by man would be exposed for the ineffectual, pittance that it truly is.

Global warming proponents claim the science is "settled". Well, if that was true, why do governments around the world continue to funnel billions of dollars into continuing research of human contributions to global warming? If the science is indeed settled, why not put this money into better use, such as, finding a cure for cancer? The premise that any science is settled is absurd. Science is always evolving. Just look at the "food pyramid"; how many times has that changed?

There is a plethora of scientific research showing that climate change has routinely occurred over decades, centuries, millennia, eras, periods, epochs and ages. Global warming theory is almost exclusively supported by disjointed, daily, anecdotal "evidence" that proves absolutely nothing, such as, four polar bears were found floating dead in the Arctic Ocean. (Here's a news flash for AGW proponents: when bears die in the wild, they do not get a formal cemetery burial, a New Orleans jazz funeral march, or have their cremated remains put into an urn)!

There is no doubt that man contributes to global warming. There is also no doubt that nature contributes as well. The question should not be "do you think man causes climate change", the question should be "how much does man contribute to climate change". As summarized in A Convenient Fabrication (ACF), man's total contribution to the greenhouse gas effect from CO2, CH4, N2O and CFCs is only 0.28% The remaining 99.72% of the greenhouse gas effect is due to natural causes including emissions from the oceans, rotting vegetation, photosynthesis, and animal respiration.

Even global warming advocates and alarmists are willing to admit that volcanic eruptions have a temporary cooling effect on the climate. But what almost all these people do not know is that volcanic eruptions are an anti-correlation with secular sunspot activity. That is: With fewer sun spots, there are more volcanic eruptions. What would shock the AGW folks is that the glacial advances of the last 40,000 years, also, coincide with volcanic eruptions (Bray J. R., 1974, Volcanism and glaciation during the past 40 Millennia, Nature **252**: 679). In other words, the fewer sun spots there are, the more glaciers advance.

Recently, claims have been made correlating an increase in eruptions of volcanoes buried under ice with increases in greenhouse gases caused by man. The theory here is that the decrease in glacier weight caused by the melting of glaciers allows eruptions in volcanoes to be released more easily. To this I say: These guys cannot be serious! The forces inside the Earth, which easily push around entire tectonic plates covering thousands of miles in area and dozens of miles in depth could care less about a few feet of ice at the surface. Next, these same "scientists" will be attributing increased sunspot activity on the Sun to the wearing of sunglasses on Earth.

"It's tough to make predictions, especially about the future"

- Yogi Berra

Chapter 22:
The Pillars of
Global Warming

There is not one piece of empirical evidence linking human activities to the climate - NOT ONE. Arguments for global warming and climate change can be grouped into four categories: anecdotes, computer projections, Hockey Sticks, and consensus. These are the pillars supporting the global warming theory. These are not proofs of global warming. They are nothing more than logical fallacies, or arguments rooted in logical fallacies.

LOGICAL FALLACIES

In the legal profession there is a saying: 'When the facts are against you, argue the process. When the process is against you, argue the facts.' When things are not going your way in an argument or debate, many times one resorts to logical fallacies. There are a couple of dozen of these little goodies, such as, the Ad Hominem, the Appeal to Authority, the Strawman, the False Cause, the Slippery Slope, the Loaded Question. Logical Fallacies are vitally important tools in warmistas tool box, because they have nothing else to back up their claims. Here are a few of their favorite logical fallacies and how they are used in the global warming climate change debate.

Ad Hominem

Ad Hominems are used to attack your opponents character or personal traits in an attempt to undermine the argument. Some of the more commonly heard Ad Hominems are: You are in the pockets of Big Oil. You are a planet basher. And the all-time favorite: You are a denier.

These Ad Hominens are rooted in the belief that the warmistas are the high priests in sole custody of climate truth. And those not complying with this belief of global warming are heathens that must be rounded up and imprisoned or reeducated by the Church of Global Warming before the non-believers can be released back into society.

The Plea To Authority

There have been statements from the heads of prestigious societies, which could lead one to believe that they represent the views of a large segment or the majority of their members. On this account, you would be wrong. I suspect the heads of these societies, being more political than technical, often overstep their office.

Statements or positions on policy have come from the American Physics Society, American Chemical Society, American Meteorological Society, and American Society of Heating Refrigerating and Air Conditioning Engineers to name a few. As a member of ASHRAE since 1986, I was profoundly disturbed when the then-President took the position on global warming and quoted out of Al Gore's book, An Inconvenient Truth. I responded to the President with a letter stating that he does not speak for me, nor probably for thousands of my fellow members, and to please keep his opinions to himself.

Probably the most egregious violation of heads of agencies overriding their flock is the IPCC itself. Quite often the conclusions reached by the various committees of scientists within the IPCC are overridden by their bureaucrat masters. An example of this was in one of the recent AR reports, in which the Executive Summary was released months before the findings of the committees were finalized. From a logical standpoint, how can a conclusion be reached when the supporting science is not yet complete? Was this meant to be coercion by the masters, telling the scientists what to furnish to support the conclusions reached?

Appeal to Emotion

This is the attempt to elicit an emotional response when lacking facts in an argument, such as, 'Don't you care about all the lives lost or disrupted with climate change?' Here an attempt is made to shift the blame for the torment being endured by victims of hurricanes or other natural disasters onto someone who obviously is not the cause, but because he is lacking of emotion on the issue should be, like Mussolini, strung up in the square.

The most recent appeal to emotion has come from a 17-year old high school student Greta "How dare you" Thunberg. In a speech to the United Nations, she claimed 'we have stolen her dreams and her childhood', 'that entire ecosystems are collapsing', 'we are at the beginning of a mass extinction', and 'causing irreversible chain reactions beyond our control'.

This appeal to emotion would have been mute if it came from boring, monotonic Al Gore. However, this appeal was thought to be so genuine coming from a defenseless little girl. Her message was considered so powerful, that she was chosen for Time magazine's 2019 Person of the Year. She was also nominated for the 2019 Nobel Peace Prize.

I do not need, nor do I deserve, to be lectured to by a little drama queen just coming out of puberty, who has no idea of how the climate works. She should take a few courses in Earth science before attempting to lecture people about the climate, who are much older and wiser.

ANECDOTES

Anecdotes are short, obscure historical or biographical accounts. Anecdotes cannot be traced to one another or anything else. Anecdotes may be headline-grabbing, emotionally charged, and tug at your heart-strings. The anecdotal myths about the appearance of polar vortices, declining sea ice extent, sea level rise, or stronger storms, droughts, floods, or forest fires, etc., simply have not materialized. Historical data shows that all these events have occurred regularly hundreds and thousands of years ago, before mankind began burning fossil fuels in earnest.

Then there are personal observations spanning a few decades, such as, "There have never been seals this far south before." Or, "There have never been armadillos north of highway umpty-ump before." Or, "We've never seen this before." Or, "We've always seen this before."

There are many websites which list hundreds of articles, each proclaiming something that is due to global warming or climate change. Just enter 'list of things caused by global warming' in any search engine and you can check them all out. Some of these articles are hilarious. The Earth is spinning slower, spinning faster, ready to explode, turn upside down. Some are contradictory, such as, there will be increases in snow and decreases snow, or lakes getting bigger and smaller. Then there are the invasions of carp, caterpillars, cats, crabgrass, herons, jellyfish, king crabs, lampreys, midges, pine beetles, and rats. Trees growing slower, growing faster; more colorful, less colorful; more lush, are in trouble. Women are cheating more on vacation. The anecdotes are endless and meaningless. Anecdotes are not proof.

COMPUTER PROJECTIONS

Computer projections is something the IPCC and the world of science stakes way too much of its reputation on. The Ludic Fallacy was identified by Nassim Nicholas Taleb in his book The Black Swan (2007). Taleb stated the Ludic Fallacy as "the misuse of games to model real-life situations".

General Circulation Models used by the IPCC

General Circulation Models or GCMs are the bread and butter of the AGW religion. This is where a lot of grant money is burned up buying supercomputer hardware and developing endless software all combining to produce projections of anything from rising sea levels to increases in temperature, all due, of course, to man made global warming.

Since so much of the Church of AGW rides on the outcomes of these GCMs, I decided to do a search on the subject. One of the first hits I got searching the Internet for GCMs was a four page paper entitled "Using The IPCC Climate Models", the IPCC's IDDRI Invulnerable Fact Sheet. If this information was from the IPCC, then it must be taken seriously. Right?

The very first sentence after "The hierarchy of models" states: "In the new IPCC AR4 report on climate change, published in 2007, several types of models are used to study how the climate changes under the impact of human activity". Did you catch that? The IPCC isn't using models to look for climate change from wherever it comes from, it is using models only to find climate change impact from humans. If there are natural impacts to climate change, the IPCC is apparently not interested in quantifying what they are. The only thing that matters is the impact from humans. Nature be damned, no matter how large that natural contribution may actually be!

The next sentence reads: "These different models and their complementarity are illustrated in Diagram 1". Before going to Diagram 1, I quickly dashed off to my 4.7-inch thick Webster's New Twentieth Century dictionary to find out what *complementarity* meant. I was shocked to find there was no listing. I panicked. I then dashed off to the Internet. Then relief. I found the answer. Complementarity is (and I hate these useless dictionary definitions which give no real additional insight) the quality of being complimentary. Well, why didn't the IPCC just say so?

Diagram 1 is labeled: "the different stages of future climate change according to the AR4 report". The first of the eight blocks in this flow chart diagram is identified as: "Socio-economic hypotheses". Socio who? What in blazes does 'socio-economic hypotheses' have to do with predicting climate change? The next block is labeled 'CO2 Emissions'. So, before any emissions are even looked at, socio-economic hypotheses have to be defined. (You really have to wonder what these guys are smoking when they put these reports together at the IPCC).

The second paragraph rambles on about how the two French models are used as inputs to other models, which ultimately encompass 23 models, concluding with: "In short, these climate models are appropriate for representing large scale climate processes while the regional models are capable of representing smaller scale processes". I beg to differ! These climate models are woefully inadequate. They fail to look beyond the end of their AGW GHG noses. They ignore effects of the moon on the oceans. They ignore the geomagnetic and solar wind influences of the Sun on our atmosphere, including N2O production, Ozone destruction, and increased UV penetration. They ignore the Impulse Of the Torque (IOT) effect of the Jovian planets on the Sun which transfers angular momentum back and forth between the Sun and planets regulating the sun spot and magnetic cycles of the Sun, which ultimate dictates our climate. They fail to account for the effect of Earth's radiation input resulting from crust/mantle friction. All these models do is keep a lot of otherwise unemployable people busy in a gigantic cocoon filled with supercomputers, studying atmospheric trace gas effects on radiation balances, oblivious to the true forces within the solar system which totally dictate our climate.

Under Model Forcing, the Fact Sheet continues: "To produce climate projections, the atmospheric concentration of greenhouse gases is modified during the simulation according to different scenarios drawn up by economists". SAY WHAT? Economists? Economists are drawing up climate change and global warming scenarios? You have got to be pulling my leg! I thought programmers under the guidance of scientists were putting all these models together. If economists are drawing up climate model scenarios, this begins to explain a lot of things.

Model forcing continues: "This modification of the atmospheric composition disturbs the radiation balance". That may be true, but there are many more modifications of the atmospheric composition which are imposed which can disturb the radiation balance, not to mention changes in temperature and pressure which affect atmospheric and oceanic circulations. There is the Sun's lower magnetic field strength represented by fewer sun spots, which results in less solar wind, which permits the entrance of more galactic cosmic rays into the atmosphere, which causes more low level clouds to form, which results in increased global cooling. There are x-ray photons which dislodge electrons from Oxygen and Nitrogen atoms creating ionic species which enter into and alter atmospheric chemistry.

So, the only forcings used are changes to greenhouse gases. No other inputs are deemed important by the IPCC. Pay no attention to the changes in the Sun's output, changes in cloud cover, or solar geomagnetic influences. Just keep your eyes on the shiny GHG object.

The Model Forcings continues: "This modification of the atmosphere composition disturbs the planet's radiation balance". Oh really! And what disturbance is involved? Care to share any insights with us? Would these disturbances be too difficult for us mortals to understand?

Model Forcing continues "Greenhouse gases (CO_2, CH_4, N_2O, O_3, and CFCs) and sulfates (SO_4) are responsible for the radiation disturbance in the scenarios. Wait a minute. Something is missing. The IPCC in its own words has stated that the most important greenhouse gas is water vapor. What happened to good old water vapor? Apparently the models do not address the influence of the most important GHG, water vapor. And the reason given is that they do not know how to account for the clouds, oceans, hydrosphere and cryosphere in their models. Oh, I see. Since they cannot account for them, they just forget about them. How convenient! Pay no attention to water vapor which has 26 times more of a spectral absorption bandwidth or Greenhouse Gas Effect (GGE) than CO_2. Nothing to see here. Just move along.

So, here we are. The elephant in the room of climate change, good old water vapor is ignored to the exclusion of CO_2. Thus, the world of consensus science stays preoccupied studying the mole hill (CO_2), while oblivious to the mountain (water vapor).

Computer Projections

Computer projections are mental gymnastics based on dubious initial conditions. The preponderance of computer projections which predicted increasing global surface air temperatures have failed. The reason: their only input is greenhouse gases.

No attention is paid to how the planets tug away at the Sun, which generates circulations on the Sun which are mirrored on the Earth. No attention is given to the Moon with its elliptical orbit around the Earth or its 18.6 year declination cycle both of which generate predictable ocean circulations on Earth. No attention is given to how cosmic rays are throttled by the Earth's magnetosphere by cyclical solar winds, causing cycles of cloud-induced global cooling. This lack of attention to the titanic forces at work within our solar system, and the relationship between our Earth, Moon and Sun are conveniently forgotten. Historical evidence of natural patterns is not given any merit. What remains, what prevails, what is adored is the Golden Calf of greenhouse gas global warming. Is it any wonder why all the computer projections endorsed and touted by the IPCC are always wrong? They are based on incorrect input and lack of relevant input.

Within the last 24 years the IPCC has issued five of its world-famous Assessment Reports, including AR1 in 1990, AR2 in 1995, AR3 in 2001, AR4 in 2007, and AR5 in 2014. These AR's are considered by establishment science as the "go to" source for everything having to do with global warming, climate change, climate catastrophe, or whatever else they are referring to today having to do with the weather. The ARs have been consistent in projecting temperature increases. The ARs have also been consistently wrong.

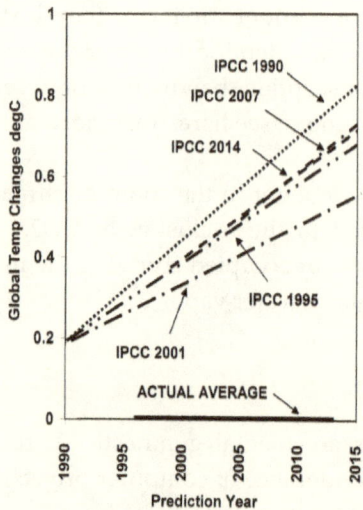

Fig 22a: IPCC AR1-5 Temperature Projections

This is the very definition of insanity. The IPCC keeps submitting the same projections over and over, but are expecting a different result, even when past results with the same projections have not panned out – when the actual satellite temperature change since 1998 has been essentially unchanged.

What is particularly disturbing is the AR5 projection which was published in 2014. Look for the 2014 projection from 2005 to 2013 in the above figure. It is buried in between and almost indistinguishable from the projections of 1995 and 2007. At this point in time the IPCC was fully aware of "the pause" in global warming. Yet with the failure of their four previous projections and the reality that there has been no global warming for the past 18 years as of September 2014, the IPCC is still compelled to provide almost the identical rates of temperature rise as previous projections. All of these projections simply have not materialized.

As you can see from the above, the IPCC is batting 0 for 5. With 5 forecasts by the IPCC, they have not been right once. Even with a flip of the coin the IPCC should have hit one or two correct projections, accidentally. But the problem is this: when you use only CO_2 or greenhouse gases as a basis for your projections of temperature - without contributions from any other sources, such as, the Sun or the oceans - the results speak for themselves. If you are only going to look at the oil pressure gauge on your dashboard to find out which direction your car is heading, you are always going to be disappointed with the navigation results.

HOCKEY STICKS

Hockey Sticks are obtained by cobbling together two unrelated climate proxy data sets. Examples of proxies include data from ice cores, tree rings, sub-fossil pollen, boreholes, corals, and lake/ocean sediments.

Hockey sticks get their name from the shape which they represent. It is a graph starting with essentially a horizontal blade of data resembling the handle of a hockey stick, which suddenly turns upward like the blade of a hockey stick. Hockey Sticks in the world of global warming are the cobbling together of two different graphs. Imagine if you will an adolescent in public school with a pair of hopefully dull scissors cutting up two different graphs and gluing them together with library paste. This activity busies the mind of the adolescent. However, this same activity conducted by adults with advanced degrees in science should be mocked and ridiculed. Instead, the Hockey Stick of Dr. Michael Mann and his fellow scissor swingers actually became the centerpiece of the IPCC's Third Assessment Report, AR3.

This now famous expression of "the hockey stick" which crept into the lexicon of global warming rhetoric describes a graph of temperature which appeared in a scientific paper by Mann, Bradley & Hughes (MBH) published in 1998. The graph is a synthesis of recent temperature records which is the blade portion of the hockey stick, and a biased collection of proxy data (ie: tree rings, corals, etc.) used to reconstruct temperatures which is the stick portion. After only a few years of being on the scene this graphic was quickly hailed as the ominous proof of global warming. Then Ross McKitrick and Stephen McIntyre (M&M)

published their critique of the Hockey Stick, in Energy & Environment, Vol 14, No 6, 2003. <www.multi-science.co.uk/mcintyre-mckitrick.pdf>

The following errors were found by M&M in MBH98:
- unjustified truncation of 3 series
- copying 1,980 values from one series onto another series, resulting in incorrect values in at least 13 series.
- displacement of 18 series to one year earlier than apparently intended.
- unjustified extrapolations or interpolations to cover missing entries in 19 series.
- geographical mislocations and missing identifiers of location
- inconsistent use of seasonal temperature data where annual date are available
- obsolete data in at least 24 series, some of which may have already been obsolete at the time of the MBH98 calculations
- listing of unused proxies
- Incorrect calculation of all 28 tree ring principal components

Kind of makes you wonder, who was paying any attention to this study during 'peer review', or did the reviewers just gave it the usual AGW blessing of anything AGW.

Peer Review

In order to get funding to do research the requestor must prove he has the resources - the manpower and materials - to accomplish the task. The requestor most show his organization has an Independent Review Board to ensure the work is being done ethically. The usual suspects involved with climate research have access to this facility. Small timers or independent researchers will have more difficulty.

If you are a writer of scientific papers without the benefit of an Independent Review Board or co-authors to assist in your endeavors, you are on your own. Your attempts to break into the status-quo are daunting. Your claims and work will likely be either ignored or ridiculed. Your peers will be pressured to view you with contempt. A scientist who goes against the inner circle can easily be "blackballed" by his fellow scientists. The scientist who does research on natural causes of global warming or climate change will quickly find himself excommunicated from the 'consensus'. He will find it difficult to obtain funding. He will be ignored by the main stream media. He will be alone.

If you are a member of the inner circle like Michael Mann, James Hansen, Phil Jones, Kevin Trenberth, Keith Briffa, Susan Solomon, Bernie Santer, Gavin Schmidt, Lonnie Thompson, et al, you are revered. You will have easy access to vast funding from numerous government agencies and private organizations. Your work will be openly accepted. Your papers will be received and given priority review for publishing. You are in the "In Crowd". If you so desire, you have a job for life grazing at the vast trough of taxpayer funding.

There is one reason why there is a preponderance of scientific papers which conclude that global warming or climate change is the fault of man: Grants are dispensed, but only if the goal is to find that mankind is the cause of global warming or climate change. Try getting a grant showing that mankind is NOT the cause – good luck with that.

Peer review can be defined as: evaluation of scientific, academic, or professional work by others working in the same field. However, more often than not, the "others" in the same field are a very small cadre all of whom are quite familiar with each others work, all with very similar agendas. It is not unusual for any member when asked to peer review a paper, can tell by the paper's subject, writing style, agenda, etc., who the author is. Assuredly, the reviewer will likely give his approval of the paper with virtually no comments, unless the subject is totally at odds with the reviewer's agenda, in which case this paper will be torn to shreds and rejected.

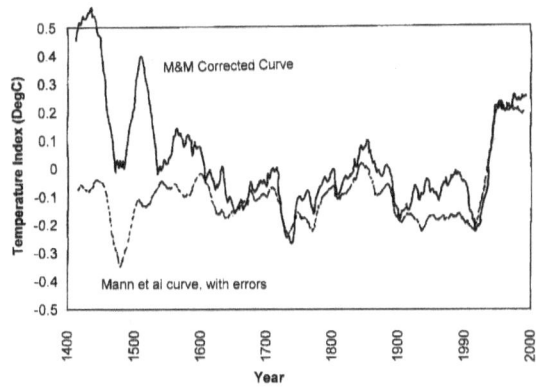

Fig 22b: Two Hockey Sticks
<www.multi-science.co.uk/mcintyre-mckitrick.pdf>

211

The dotted curve in the graph above is the original Mann et al Hockey Stick with errors. The solid curve above is the M&M corrected version of the Hockey Stick. Great effort was made by Mann et al to hide the Medieval Warming Period, to give the appearance of a sudden temperature rise.

What is amazing about Dr. Mann is his rationale not to release his data gathering methodologies and statistical algorithms, so that his peers could reproduce the same findings. Dr. Mann's reluctance was his belief that all the other scientists wanted to do was to prove him wrong. Well, Dr. Mann, welcome to the real world of science that demands replication of results by others, not results that only standup in your cacooned little fantasy world. The rest of us are not fooled by a theory constructed on the analysis of a single bristlecone pine, YAD061, from the Yamal Peninsula, with the rest of the world's pine cone bristles told they can go take a hike. Anyway, by the time the next AR5 came out, any mention of the now discredited Hockey Stick all but disappeared.

These Hockey Sticks or, as I like to call them, FrankenGraphs, would have received an "F" in Junior High School science class 50 years ago. And yet, incredibly, these graphs-of-the-imagination are still embraced by many scientists today. Hockey Sticks are artificial fabrications. They are not proof.

CONSENSUS

Consensus is an opinion or position reached by a group as a whole. Millennia and centuries ago the 'consensus' believed the Earth was the center of the Universe. With the exception of the members of "flat earth societies", or followers of religious edicts, hopefully everyone else today knows the Earth is revolving around the Sun, not the other way around. Consensus is just another word for the Appeal to the Masses, Appeal to Belief, Appeal to Democracy, Appeal to Popularity, Authority of the Masses, or Bandwagon Logical Fallacy. Popularity of an idea is no guarantee of its validity. Consensus is not proof.

This brings up a sideline thought, which has two possibilities. Either the Warmistas are getting desperate when a study is commissioned to find out what a consensus has to say, or the grant money machine really doesn't care what is being studied so long as the conclusion reached is that man and not nature is responsible for climate change.

I recall the Reverend Al Sharpton appearing on CNN, chiming-in on the issue of manmade global warming. (You know the argument about mankind being responsible for global warming is being lost when proponents have to march out Al Sharpton in a white lab coat. The only thing he was missing was a stethoscope. Is this all the Warmistas have?).

The 'Ninety Whatever Percent of Scientists' Hoax

Recently, there have been numerous studies proclaiming that 90%, 93%, 97%, 98%, ninety-whatever percent of scientists agree that man is responsible for climate change. This all started with a Naomi Oreskes study in 2005. It was followed by studies by Doran and Zimmerman in 2009, Anderegg in 2010, Cook in 2013, and others. All these studies had one thing in common. They didn't provide proof that mankind was responsible for global warming or climate change. They provided a plethora of studies by scientists who agreed that mankind was the cause.

Oreskes

A University of California at San Diego scientist, Dr. Naomi Oreskes, published in Science magazine a massive study of every peer-reviewed science journal article on global warming from the previous 10 years.

(Naomi Oreskes, The Scientific Consensus on Climate Change, Science, 03 Dec 2004: Vol. 306, Issue 5702, pp. 1686)

Oreskes did not study every peer-reviewed article from the previous 10 years. Oreskes' study was based on an database search using the three keywords "global climate change", which resulted in her eventual 928 "hits". Had she used only the two keywords "climate change" she would have gotten almost 12,000 "hits". This oversight, failed to pick up the 11,000 or so other peer-reviewed research papers which show that global temperatures were similar or even higher during the Holocene Period and the Medieval Warm Period when CO2 levels were much lower than today; that solar variability is a key driver of recent climate change; and that climate modeling is highly uncertain. According to Naomi Oreskes who prepared the study, "75% of the abstracts 'either explicitly or implicitly accept the consensus view" that man is the cause of global warming. (Science, 2004). Ultimately, on December 15, 2004 Dr. Oreskes admitted there was a serious mistake in her Science essay.

Dr. Benny Peiser of Liverpool John Moore University using the same three keywords came up with 1,247 documents of which 1,117 had abstracts. The results of his analysis contradict Oreskes' findings and essentially falsify her findings as follows:

- 13 (1.1%) explicitly endorse the 'consensus' view.
- 322 (28.8%) implicitly endorse the 'consensus' view
- 89 (8.0%) focus on "mitigation"
- 67 (6.0%) focus on methodological questions
- 87(7.8%) are unrelated (deal with paleo-climatological research)
- 34 (3.0%) doubt/reject the last 50 years of warming as manmade
- 44 (3.9%) focus on natural factors
- 470 (or about 42% of the balance) do not include any direct or indirect link or reference to human activities.
- there are 2.5 times as many abstracts that are skeptical of the notion of manmade global warming than those that explicitly endorse it.
- Science should withdraw Oreskes' study and its results in order to prevent any further damage to the integrity of science.

Dr. Peiser submitted his findings to Science magazine, but were rejected on the basis that his results were "widely dispersed on the internet". Science also took exception to Peiser's findings as being "not perceived to be novel" (Matthews, 2005). Is there any doubt had Dr. Peiser's findings endorsed the consensus view he would have been published?

Doran and Zimmerman

The study that generated the "97% of scientists" nonsense was P. T. Doran and M. K. Zimmerman, Examining the Scientific Concensus on Climate Change, EOS Trans AGU, 90(3),22. The researchers involved should also be a red flag - a college professor and his graduate thesis student.

This study involved an online survey sent to 10,257 Earth scientists, which generated 3,146 responses. Of these responses, 77 were from climate scientists. Of these 77 climate scientists, 75 answered 'Yes' to

Question #2: 'Do you think human activity is a significant contributing factor in changing mean global temperatures?' Therefore, in the minds of Doran and Zimmerman, 75 divided by 77 equals 97%. Or does it?

There were nine questions asked in the Doran & Zimmerman On-line questionnaire, as follows:

Q1. When compared with pre-1800's levels, do you think that mean global temperatures have:
1. Risen
2. Fallen
3. Remained relatively constant
4. No opinion/Don't know

First, if any scientist is aware of the GISP2 Ice Core Data compiled by IPCC Contributor Richard Alley PhD the only possible answer is "2. Fallen".

Second, if any scientist answers anything else but "2 Fallen", his survey should be tossed into the dumpster, because he/she is ignorant of this important, irrefutable, historical evidence and does not know what he/she/it is talking about.

Third, "do you think" is a speculation. Either the scientist knows, or does not know. Either the scientist has evidence, or does not. I am not interested in what a scientist, or anyone else 'thinks'. I want - we should all demand – evidence, not useless speculation.

Q2. Do you think human activity is a significant contributing factor in changing mean global temperatures? [This question wasn't asked if they answered "remained relatively constant" to Q1]
1. Yes
2. No
3. I'm not sure

First, again, there's that 'do you think' speculation. If D&Z were intellectually honest, the question asked should have been 'can you quantify the impact of human activity on the environment?'. If, the answer to this question was 'no', then the answer from this scientist should, again, be thrown into the dumpster.

Second, just what is a 'significant' contributing factor? If D&Z were seeking truth, the question asked should have been: in percent, what is mankind's contribution to changing global mean temperature? If this answer was left blank or answered 'do not know', it's dumpster time, again.

Third, the correct answer is this: Only 3.27% of all CO2 generated comes from man, the other 96.73% comes from nature. If man ceased to exist, the reduction in the GGE, 1 part out of 357, would be barely noticeable.

> Q3. What do you consider to be the most compelling argument that supports your previous answer (or, for those who were unsure, why were they unsure)? [This question wasn't asked if they answered "remained relatively constant" to Q1]

There are just too many definitions for an argument. Is it an exchange of views, is it a reason(s), or is it something else? The question that should have been asked is: What *evidence* do you have that supports your previous answer.

> Q4. Please estimate the percentage of your fellow geoscientists who think human activity is a contributing factor to global climate change.

First, what a worthless question. Why would any scientist care what other scientists are thinking about. This isn't a popularity contest, looking around the classroom at fellow students for raised hands before you decide to raise your hand, or not, for teacher to see. Galileo and Copernicus didn't care what their fellow scientists 'thought'. And just what criteria should the scientist use to 'estimate' this percentage?

Second, why is it suddenly important to *quantitatively* estimate the number of scientists who think about something, but are satisfied to only *qualitatively* estimate mankind's impact on global temperature? Do you think D&Z's objectives are misplaced here?

> Q5. Which percentage of your papers published in peer-reviewed journals in the last 5 years have been on the subject of climate change?

The answer to this would be valuable in estimating how much time any one scientist spends grazing at the trough of free taxpayer money to do research, research that would only be funded if the conclusion reached is that man has an impact on the climate.

Q6. Age

Really, should we care about the age of any one scientist? There are bright young geniuses, and senile old codgers. Of what possible measure is age important? Is older better than younger? And, if so, how so?

Q7. Gender

This question at best is irrelevant and at worst is chauvinistic.

Q8. What is the highest level of education you have attained?

Sounds to me like BS, MS, PhD; Bull ****, More ****, Piled higher and Deeper.

Q9. Which category best describes your area of expertise?

This is a fair question. It would seem more worthwhile in knowing the opinion of a climatologist rather than the opinion of a proctologist when it comes to the climate. But knowing a category does not guarantee the expertise of the scientist. Knowing how many times his/her predictions have come true in the past might be interesting, but – as they say on Wall Street – this does not guarantee future results.

So, let's recap. Surveys are carefully worded to elicit the responses desired. Question 1 could have asked "compared to the last 500,000,000 years, do you think the Earth is warmer today"? Anyone who has read the CO2 Record chapter of this book or is familiar with any of the supporting literature could not come to any other conclusion other than *No!* Question 2 could have asked "Do you think nature or man is the primary driving force behind climate change?" Again, if anyone had read the CO2 Record chapter of this book or is familiar with any of the supporting literature, the only conclusion that could be reached is: *nature is the primary driving force.*

Of the 77 respondents who identified themselves as climate scientists, 75 who had more than half of their papers accepted by peer-reviewed journals answered *Yes* to Question 2. Thus, using the math of Doran and Zimmerman, 75 divided by 77 equals 97%. By my calculations 75 divided by 3,146 respondents equals 2.38%. What percentage do you think it should be? And what does it matter, anyway?

Given that there are tens of thousands or even hundreds of thousands of scientists with real expertise in basic sciences related to climate, a survey that looks at the views of only 77 climate scientists who had more than half their papers accepted by peer-reviewed journals is ridiculous. Its tiny sample size makes any results reached statistically meaningless.

At issue is not whether the climate warmed since the Little Ice Age or whether there is a human impact on climate, but whether the warming is unusual in rate or magnitude; whether that part of it attributable to human causes is likely to be beneficial or harmful and by how much; and whether the benefits of reducing the human contribution will outweigh the costs, so as to justify public policies aimed at reducing it. The survey is silent on these questions.

The survey by Doran and Zimmerman, also, fails to produce evidence that would back up claims that "scientific consensus" about the causes or consequences of climate change. They simply asked the wrong question. And the "97 percent" figure so often attributed to their survey refers to the opinions of only 77 climate scientists, which is nowhere near a representative, statistical sample of scientific opinion.

Anderegg et al

The "98% of scientists agree man is responsible or global warming' comes from a second study (Anderegg et al., "Expert credibility in climate change," in the Proceedings of the National Academies of Sciences: http://www.pnas.ora/content%earlypnas.ora contentleark/2010/06/04/1003187107).

To paraphrase the abstract of the above study, 'Here, we use an extensive dataset of 1,372 climate researchers and their publication and citation data to show that (i) 97-98% of the climate researchers most actively publishing in the field support the tenets of ACC outlined by the Intergovernmental Panel on Climate Change, and (ii) the relative climate expertise and scientific prominence of the researchers unconvinced of ACC are substantially below that of the convinced researchers.'

Note that this is not a survey of scientists, whether "all scientists" or specifically climate scientists. Instead, Anderegg et al. counted the number of articles published in academic journals by 908 "climate researchers," defined as people who had signed petitions opposing or supporting the IPCC's positions or had coauthored IPCC reports and had published a minimum of 20 climate publications.

They found that 97 to 98 percent of the most prolific 200 climate researchers, so defined, appeared to believe that "anthropogenic greenhouse gases have been responsible for 'most' of the 'unequivocal' warming of the Earth's average global temperature over the second half of the 20th century."

Observe that this counting exercise did not determine how many of these authors believe global warming is a crisis, or that the science is sufficiently established to be the basis for public policy, or even that future global warming would be bad (or good). Anyone who cites this study in defense of these views is mistaken.

Anderegg et al also didn't count as "skeptics" the scientists whose work exposes gaps in the man-made global warming theory or contradicts claims that climate change will be catastrophic. Dennis Avery identified several hundred scientists who fall into this category, even though some profess to still "believe" in global warming.

Looking past the flashy "97-98%" claim by Anderegg et al., you will see the study found the average skeptic has been published about half as frequently as the average alarmist – 60 versus 119 articles. Most of this difference was driven by the hyper-productivity of a handful of alarmist climate scientists – the 50 most prolific alarmists were published an average of 408 times, versus only 89 times for the skeptics.

So what, exactly, did Anderegg et al. really discover? That a small clique of climate alarmists got their writing published hundreds of times in academic journals, something that probably would have been impossible just a decade or two ago. Anderegg et al. simply assert that those "top 50" are more credible than scientists who publish less, but Anderegg, et al make no effort to prove this.

Cook et al

The Cook et al study (Cook, J et al. 'Quantifying the consensus on anthropogenic global warming in the scientific literature.', Environmental Research Letters 2013; 8 024024) analyzed 11,944 abstracts from scientific papers.

First of all, I have a problem with the basic premise of this study, which is trying to quantify a consensus. Who really cares how many scientists are agreeing with each other. The AGW/CC industry travels around in a pack mentality, like the medical and educational industries. It is not unheard of for these industries after following decades of their own conclusions to radically change course and establish another position. In the process, no explanation is given for the radical change, and no apology is made for the last few decades of their position being wrong. The change just happens with no explanation, hoping you will not notice the radical new position.

The Cook study brakes down the abstracts into the following seven categories:

1. Explicitly endorses and quantifies AGW as more than 50%: 64
2. Explicitly endorses but does not quantify or minimize: 922
3. Implicitly endorses AGW without minimizing it: 2910
4. No Position: 7970
5. Implicitly minimizes/rejects AGW: 54
6. Explicitly minimizes/rejects AGW but does not quantify: 15
7. Explicitly minimizes/rejects AGW as less than 50%: 9

The first three categories of scientists above, either explicitly or implicitly endorsing AGW, total 3,896. The 'no position' camp of scientists account for 7,970. The remaining scientists who reject AGW total 78. So, where does the 97% of scientists come from? It comes from dividing all endorsees (3,896) by all endorsees plus all rejectees (3,896 + 78) = 97.48%. This statistical malfeasance totally ignores the 'no position' category and assumes the total sample size is 11,944 minus 7,970. If done correctly, the true number of scientists who endorse AGW is 3,896 divided by 11,944 or 32.6%. So much for the '97% of scientists agree that man is the cause of global warming' nonsense.

Summary

Meanwhile the only true "proof" of global warming from the alarmists is the nonsense and lies coming out of the IPCC. The entire basis of the IPCC is its projections, based on global climate models, or GCM's. The fact is: these models are wrong and have never produced the projections forecasted. But rather than rework the models to include new inputs, the die-hards keep tinkering with the data in hopes that they somehow can make their primitive, ineffective models provide the conclusion they want. This technique is called "dry lab-ing" and is cheating and fraudulent.

The development of anything new in science starts with a hypothesis. From this, logic and previous knowledge is used to formulate a theory. Eventually, after enough testing and experimentation is performed the theory may be accepted and may become law. In the case of global warming, it is based on one premise: that CO_2 increases cause temperature increases. Yet, it can be shown from numerous standpoints that this theory does not hold up to scrutiny for the following reasons:

- The GISP2 ice core data shows that CO_2 increases happen about 800 years AFTER temperature increases (Monnin, E., Indermühle, A., Dällenbach, A., Flückiger, J, Stauffer, B., Stocker, T.F., Raynaud, D. and Barnola, J.-M. 2001. Atmospheric CO_2 concentrations over the last glacial termination. *Science* 291: 112-114.) CO_2 doesn't cause the atmospheric warming, it is the ocean warming that drives CO_2 out of solution from the oceans into the atmosphere.

All it takes to disprove a theory is ONE fact that does not support the theory. The above fact does this. Want some more facts?

- 550,000,000 years ago CO_2 levels were about 7,000-ppm, 18 times higher than today. (Berner 2002)
- For the last 600,000,000 years temperatures have varied between 12C and 22C. Right now we are at 14.5C, just 25% off the bottom of this range. (Scotese 2002)
- There has been absolutely no correlation between CO_2 and temperature within the last 550,000,000 years. (Berner and Scotese)

- Since 1998, while CO_2 has been steadily increasing, the temperature of the world's atmosphere has not measurably changed at all according to satellite measurements.
- For the last 120 years, temperatures have oscillated up and down in five cycles in perfect unison with total solar irradiance, not CO_2 concentration.
- For the last 150 years, while fuel consumption has increased, the temperature anomaly has increased three times and decreased twice.

How many facts do you need to prove that the global warming theory based on CO_2 is unsupportable?

Global warming theory is supported by four things. One, is meaningless anecdotes (eg: we have never seen armadillos on this side of highway umpty-ump before), which are traceable to nothing. Two, computer projections which could be drawn by hand with the same accuracy using the same questionable initial conditions. Three, Hockey Sticks, which cobble together two questionable proxy data sets into a FrankenGraph to arrive at a new mythical curve not representing anything, anywhere in nature. Four, consensus, which 500 years ago said the Earth was the center of the universe.

In the real world of science, the theory is adjusted to match the facts. In the global warming world of science, the facts are continually adjusted to fit the theory. The concensus is determined to pile drive the big round peg into the little square hole no matter what it takes. And it does so with about one thousand times the funding that skeptics get to do their research.

"I was utterly disgusted. My second thought was that it was inevitable. It was bound to happen. Science, not so very long ago, pre-1960s, was largely vocational. Back when I was young, I didn't want to do anything else other than be a scientist. They're not like that nowadays. They don't give a damn. They go to these massive, mass-produced universities and churn them out. They say: 'Science is a good career. You can get a job for life doing government work.' That's no way to do science."

James Lovelock, founder of the Gaia hypothesis

Chapter 23: IPCC Gates

It is hard to believe the quote above came from the founder of the Gaia hypothesis. This hypothesis assumes the world is one big organism. And within this world is a population of mean, nasty humans that are a cancer to this organism.

I wonder how Dr. James Lovelock feels about the scientists involved with the Intergovernmental Panel on Climate Change (IPCC) today? Any scientist, whether they support or criticize the workings of the IPCC must feel very sorry for the IPCC by now. Though not well publicized by the lame stream media, the IPCC has been stepping into crap with incredible consistency. And its all its own doing.

The IPCC was "established by the United Nations Environment Programme (UNEP) and the World Meteorological Organization (WMO) to provide the world a clear scientific view on the current state of climate change and its potential environmental and socio-economic consequences." I just love this socio-economic baloney. Don't you? And who founded the UNEP, none other than Maurice Strong.

Maurice Strong

Who is Maurice Strong? Hardly anyone knows the name. He came out of depression poverty to become one of the world's most influential figures. He was a businessman, stock market manipulator, mentor to Al Gore, silent partner of Generation Investment Management (Al Gore's mutual fund); acknowledged Socialist (probable Communist); ex-Canadian. He is considered by many as the father of sustainability, global warming, the UNEP, the IPCC, and carbon trading exchanges. Maurice Strong was apolitical; rubbing shoulders as easily with Barrack Obama as he did with George H. W. Bush.

Why haven't you heard of him? He elected to stay behind the scenes and just pull the strings of power at the UNEP and IPCC. Though born and raised in Canada, Strong resided in Bejing China, and for two good reasons. First, Maurice Strong's life-long dream was the destruction of the United States and making China the world's most powerful country. Second, Canada had an outstanding warrant out on him and had many questions to ask of Strong about his shady financial dealings. Maurice Strong died in his beloved Beijing China in 2015.

IPCC Projections

Looking at the curve below, the dotted line is the IPCC Best 2.0 estimate of temperature for the next 20 years made initially in 1990. The solid line is the official temperature as maintained by UAH Hadcrut. Between 1990 and 2002 it looked like the Hadcrut temperature was doing its best to keep up with the IPCC projection. Between 2002 and 2006 the actual temperature leveled off, departing from the IPCC projection. Beginning about 2006 the actual temperature started dropping, while the IPCC projection went along on its merry linear way upwards. This departure of reality from the IPCC projections is very disconcerting to the AGW scientists. Many of the usual suspects (eg: Trenberth, Mann, Jones, et al) are at a loss to explain why the CO2 keeps increasing, but the temperature since 1998 has not followed at all. Some of these scientists claim the missing heat, which was originally predicted to manifest itself in the upper equatorial atmosphere but hasn't, instead might be hiding in the deep oceans. If it isn't eventually proved to be in the deep oceans, where next will they ask us to believe it is hiding? Their efforts to keep banging away at that square peg trying to get it into that round hole seems to be never-ending.

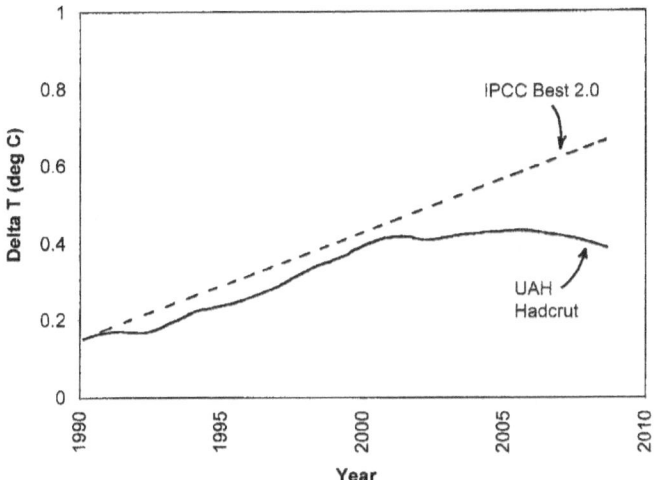

Fig 23a, IPCC Temperature Projection of 1990
and Actual Measurements, 1975 to 2012
(Based on IPCC AR Reports)

The premier findings of the IPCC are its Assessment Reports, which are issued about every five years. These are the "go to" sources on all issues having to do with climate change for government and business leaders. IPCC Assessment Reports are considered the Gold Standard of references having to do with where the climate is heading.

Just about every conclusion and recommendation developed by the IPCC is based on projections from computer models. These rudimentary models started as weather models and have progressed, though slowly and inadequately, to encompass what they believe are models of the climate. Nothing could be further from the truth. Everything in these models is stacked up to support the conclusion that CO2 – and CO2 alone – is responsible for every change in climate, floods, droughts, dust storms, fish kills, coral bleaching, earthquakes, tsunamis, plagues - you name it. Acknowledgement and analysis of historic data is not only discouraged, it is anathema.

However, for the last few years, with increasingly regularity, the IPCC has been "stepping into it". My opinion is: it couldn't happen to a finer bunch. Thus, the reason for the chapter title of IPCC Gates. Here is a sampling of the various gates the IPCC has slammed its nose into.

Temperature-Gate

One of the big arguments powering the theory of AGW, is the temperature projections that come out of the IPCC. There has been a projection in each of the last four Assessment Reports. The big problem with these projections is that they are always wrong. But just like the big story on Action News that proves to be wrong and there is no retraction, the last IPCC projection is conveniently forgotten and the emphasis is focused on the next projection using the latest whatever.

Satellite-Gate

As if it wasn't bad enough that the global surface temperature history has been turned into garbage, the satellite record of temperatures maintained by NOAA has suddenly come under suspicion. The issue started with reports of the Northern Lake Michigan Surface Temperature satellite images showing thousands of temperatures falling into the impossible range of 415F to 604F. This data was reported by the University of Michigan's Coastal Watch web site, officially connected to NOAA. <http://www.climatechangefraud.com/climate-reports/7479-us-government-in-massive-new-global-warming-scandal> But, it gets worse. These reporting problems began with satellite NOAA-16, but satellites NOAA-17 and NOAA-18, have come under scrutiny for having similar problems.

The crime here is that this obviously impossible and wrong temperature data is automatically downloaded by customers and used in their temperature compilations. Worse, NOAA knew there was a problem, but instead of warning clients of the corrupted data they swept the problem under the rug. This is your tax dollars at work in the Federal Government.

The Indian Government back in 2004 knew of these problems which were initially published by Researcher Devendra Singh. These temperature errors have been going on for 6 years with the full knowledge of NOAA, which could have warned users of the problem. Instead, NOAA decided to continued to cover-up their rear-ends and tell no one of the problem. <http://canadafreepress.com/index.pp/article/26758>

According to Kevin Trenberth of NCAR, half the heat buildup from global warming is missing <http://global-warming.accuweather.com/>. Really! So how does Trenberth know that half the heat is missing? And how is Trenberth so darn sure that the missing heat is building up in the deep oceans and will eventually show itself? It is a thermodynamic fact that heat rises. But in the world of Kevin Trenberth heat is more likely to hide in the deep oceans below colder more dense water.

Well, it now appears it showed up as erroneous data transmitted by the NOAA-16/17/18 satellites. So, now we have corrupted surface temperature measurements and corrupted satellite temperature measurements. Take your pick at which garbage you want to use as input into the Global Climate Models everyone is staking the future of temperature measurement on.

Glacier-Gate

The story begins with the assertion in the 2007 IPCC AR4 Report that the Himalayan Glaciers would melt by 2035. It turns out this assertion was based on a geography student's Masters dissertation that quoted the observations of mountain guides in the Alps, and an article in a mountaineering magazine based on anecdotal evidence from mountaineers about the changes they were witnessing in the mountainsides around them. <http://www.telegraph.co.uk/earth/environment/climatechange/7111525/UN-climate-change...>

The coordinating lead author of the IPCC report said "It related to several countries in this region and their water sources. We thought that if we can highlight it, it will impact policy makers and politicians and encourage them to take some concrete action". Oh, I see. It doesn't matter that the claim for glaciers melting is based on anecdotal reports from inexperienced individuals with absolutely no scientific credentials.

The important thing is that a crisis be manufactured, so politicians can fleece unsuspecting taxpayers to solve a problem that does not exist.

The IPCC's own mission statement requires that its reports should be neutral with respect to policy. So, why hasn't this lead author been taken out to the wood shed? Because, the policymakers at the IPCC condone this misconduct - it fits the agenda.

Sea-Level Gate

By now almost everyone has heard from one source or another, that the world's oceans are supposedly rising due to melting glaciers caused by global warming. Claims have been made that the Maldives are sinking, that the Tulvala islands are sinking, that Vanuatu is sinking, that the world's oceans are rising at the rate of 2.3-mm/century. All these claims are groundless.

Dr. Nils-Axel Mörner is perhaps the world's most renowned expert on changing sea levels. In an interview he said that a sudden 20 cm change did occur in the Maldives in the 1970's for as yet unexplained reasons, none of which had to do with sea level change. Tuvalu, which is pictured on pages 186-187 in Al Gore's book An Inconvenient Truth, gives the impression that this island is also about to be inundated by rising sea levels. This, according to Dr. Mörner, is not corroborated with the tidal gauge there that shows absolutely no change in sea level for the last 30 years.

Vanuatu is another island claimed to be sinking and in need of evacuation, but the tidal gauge there shows no level increase at all. (Interview with Dr. Nils-Axel Mörner, Claim That Sea Level Is Rising Is a Total Fraud, **EIR**, June 22, 2007). Dr. Mörner has stated that he was astonished to find out that not one of the 22 contributing authors on sea levels in the last two IPCC reports was a sea level specialist. <http://www.liveleak.com/view?i=78f_1238357641&c=1> Why would you be surprised?

What almost everyone doesn't know, is that all the hubbub about sea levels rising is based, incredibly, on the measurements taken from one – yes ONE - tidal gauge in Hong Kong Harbor. This gauge is one of 290 that are monitored by The Global Sea Level Observing System (GLOSS) operated by the British Oceanographic Data Center. <http://www.bodc.ac.uk/data>

Fig 23b: Quarry Bay Tidal Gauge Station
<http://www.hko.gov.hk/wxinfo/news/2004/Fig.2_2000_QUB2.jpg>

This one tidal gauge, identified in the GLOSS Station Handbook as Quarry Bay Station, GLOSS 77, has the enviable distinction of having recorded the now infamous 2.3 mm/century rise in sea levels. There is also something unique about this station which the IPCC, Al Gore, and global warming alarmists either do not know, or more likely do not want you to know. GLOSS 77 is situated on geology that is subsiding. In other words, this particular tidal gauge station isn't recording sea levels rising, it is recording the land it sits on sinking into the sea!

Yet, incredibly, the IPCC has hinged its entire argument of rising sea levels by using this one tidal gauge as the driving force in its computer models. All the claims of rising seal levels are not based on actual observations from all instruments available around the globe, but on computer model projections, programmed to look at one tidal gauge in Hong Kong Harbor without paying attention to the other 289 tidal gauges around the world, to produce the sensationalistic forecast demanded of the IPCC.

This cherry-picking approach used above for developing sea level rise was also used to develop the blade of the now famous Dr. Michael Mann Hockey Stick. Data from one pine cone bristle sample, YAD061, out of several Yamal Peninsula pine cone bristles was used. Not pine cone bristles from all around the world, but one pine cone bristle in one location on a peninsula jotting out into the Arctic Ocean from Russia. This shows you how desperate scientists with an agenda have become to prove a point.

Making things worse is the discovery of the University of Colorado's Sea Level Research Group's decision to add 0.3 mm every year to its actual measurements to account for "glacial isostatic adjustment" or GIA. <www.foxnews.com/scitech/2011/06/17/research-center-under-fire-for>. In public school we used to call this "a fudge factor". Perhaps this adjustment is well deserved and honest. Maybe it isn't. The fact that this kind of adjustment goes on behind the scenes, without any disclosure up front, is disturbing.

Perhaps the most egregious scientific malfeasance regarding sea level measurement was reported in Earth Systems and Environment, December 2017, 1:18. In the article, Is the Sea Level Stable at Aden, Yemen, scientists Dr. Albert Parker and Dr. Clifford Ollier seem to have uncovered shenanigans going on at the Permanent Service for Mean Sea Level or PSMSL. PSMSL is responsible for the collection, publication, analysis and interpretation of sea level data, including GLOSS. The scientists at PSMSL look like they have adopted the same scientific procedure used by Dr. James "Thumbs On The Temperature Scale" (Totts) Hansen. PSMSL, by tossing away earlier historical records of lower sea level measurement, have changed the resulting trend from no change at all or a falling trend, to a sea level rise. Unlike Hansen, who supposedly was 'homogenizing' the temperature data sets, PSMSL appears to have adjusted the data in a "highly questionable" and "suspicious" manner. PSMSL has used the cobbling of data sets reminiscent of Dr. Michael Mann, the creator of the discredited "Hockey Stick" temperature warming graph, to recreate and adjust the data to show a rise in sea level when, indeed, none exists.

Hurricane-Gate

The most compelling case for the supposed fact of increasing hurricanes was made by Hurricane Katrina in 2005. Katrina was an intense hurricane to be sure. But Katrina was a flooding event caused by failure of levies, not a wind-driven destruction event as is usually the case with hurricanes. Nevertheless, Katrina became the poster hurricane in Al

Gore's book and movie, and for months of news coverage by the MSM. Katrina also became a springboard for what some claimed as government inaction, lack of coordination of government agencies, and who was to blame for the destruction, more than it was about a hurricane.

Katrina and other anecdotal hurricane information aside, the true total global effect of all hurricanes (in the west) and cyclones (in the east), and similar storms in the Southern Hemisphere can be measured accurately by their total Accumulated Cyclone Energy, or ACE. Below is a plot of the ACE for the last 50 years.

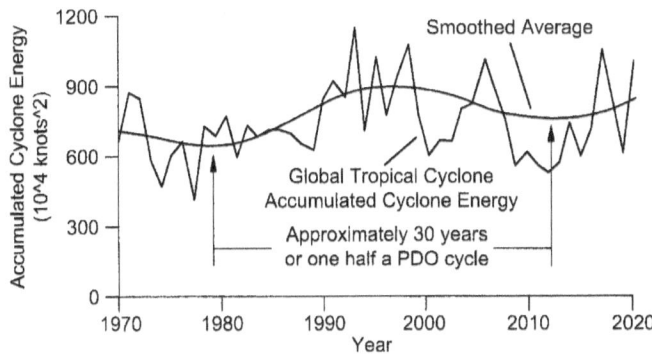

Fig 23c: 50 Year Global Tropical Cyclone Accumulated Cyclone Energy
(Based on Dr. Ryan N. Maue website climatlas.com)

From the above graph it appears a 30 year span is apparent between the two low portions of the smoothed average.

Below is a plot of the tropical storm and hurricane frequency within the last 50 years.

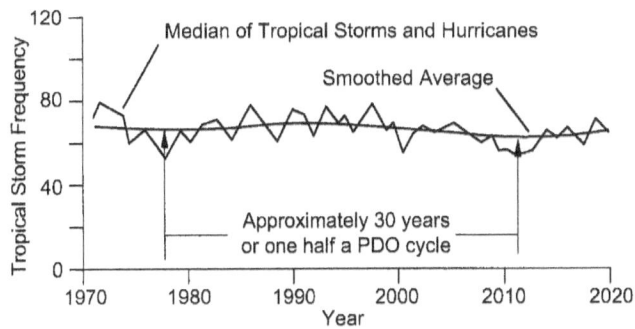

Fig 23d: 50 Year Tropical Storm and Hurricane Frequency
(Based on Dr. Ryan N. Maue website climatlas.com)

So, from the two plots above can any conclusions be drawn? Warmer oceans should be the breeding ground for more energetic storms. Is the signature of a 30 year harmonic of the 60-year Pacific Decadel Oscillation apparent? But where is the warming coming from? Changes in ocean albedo? And remember, the warming comes first, followed by the CO_2 increases.

Of course, this didn't stop the IPCC from 'stepping in it' again. With their 2007 Fourth Assessment Report (AR4). In AR4 SPM they make predictions of more intense hurricanes, and that there have been more tropical cyclones since 1970 due to human influence. Hurricane expert Chris Lansea who contributed to AR2 and AR3, got so disgusted with the politicization of the IPCC process on this issue that he resigned in 2005. <http://sites .google.com/site/globalwarmingquestions/ar4hur>

As usual the problem with the IPCC's predictions is that they are based on computer models. Well, did the models detect a 30 year signature in the Global Tropical Cyclone data? Of course they didn't, because the computer models are programmed to produced the results intended, based on cherry-picked papers by Emanuel and Webster, and ignoring papers by Pelke, a bulletin of the AMS written by 11 meteorological experts <sites google>, and the life work of Dr. Ryan N. Maue as depicted in the graphs above. If all the information had been included, the models would have likely predicted no change at all, which is what the data suggests.

Amazon-Gate

In the IPCC AR4, Chapter 13, the claim is made that "up to 40% of the Amazonian forests *could* react drastically to even a slight reduction in precipitation. *Could*? *Not will. Not would. But could*! This means that the tropical vegetation, hydrology and climate system in South America could change very rapidly to another steady state, not necessarily producing gradual changes between the current and future situation (Rowell and Moore, 2000). It is more probable that forests will be replaced by ecosystems that have more resistance to multiple stresses caused by temperature, droughts, fires, such as, tropical savannas." There are many problems with the above assertion.

First, there is no evidence given as to where this "40%" comes from. Second, the reference sighted is a non-peer-reviewed report, which violates the IPCC's own rules and, as such, can only be cited as a secondary, not primary, source. Third, the report is from an advocacy group of the WWF and IUCN, hardly an open and objective source of information. Fourth, of the two "expert" authors who wrote the report, one is a policy analyst and the other is a freelance journalist, neither of whom have scientific credentials <http://wattsupwiththat.com/2010/01/25/de-jour-gate-flaver-amazon/>. Fifth, the speculation is made that a "slight" reduction _can_ lead to "rapid" changeover from one steady-state to another. Can? But, do they? Climate systems are complicated, not fully understood mechanisms. To make this statement without full knowledge of the ecosystem dynamics (inputs, outputs, and characteristics) is pure speculation, without any basis whatsoever. Sixth, the IPCC has purposely ignored changes in precipitation and droughts as having a direct connection to ocean oscillations (ENSO and PDO), but instead makes a groundless, untraceable connection to higher temperatures of supposed manmade global warming.

Africa-Gate

The IPCC AR4 has made the claim that food yields from African countries _could_ be reduced 50% by 2020, which _could_ result in the death, of millions from starvation. _Could_? Here we go again. This claim was made in a "report produced by a Canadian advocacy group, written by an obscure Morrocan academic, Ali Agoumi, who specializes in carbon trading, citing three references, most all of which do not support his claims. The exception was the Morrocan reference, which said in serious drought years cereal production _might_ be reduced 50%. (Might? Does it ever end?) However, the Algerian reference said agricultural production will more than double. Not used were two reports prepared by the Working Group's Co-Chair which came to the conclusion that food changes would be insignificant. Also, overlooked by the IPCC was peer-reviewed work done by Dr. Mike Hulme and Dutch and German teams. <http://www.telegraph.co.uk/comment/columnists/christopherbooker/7231386/African-crops-yield-another-catastrophe-for-the-IPCC.html>

The only conclusion that can be drawn here is that the research of one individual with dubious credentials, who would likely benefit from carbon trading, was used by the IPCC to perpetuate the global warming myth. Not so incredibly, reports prepared by the Working Groups Co-Chair, which showed no significant food changes, were not used.

I cannot get enough of those photographs of those cute, cuddly polar bears perched precariously on the smallest of ice floes, seemingly holding on for dear life to survive the ravages of supposed melting ice from increasing temperatures caused by manmade global warming. The anecdotal stories and photos are relentless, all designed to tug away at your heart strings. The facts are something else entirely. The polar bears are not in trouble, their overall population has increased substantially in the last 50+ years, they have thrived in warmer climates, and in geological periods have endured much warmer climates in the past and survived.

There are 19 polar bear populations encircling the Arctic Ocean. Of these 19 populations, 45% are stable, 25% are unknown, 16% are decreasing, and 14% are increasing (Norris et al, Polar Bears At Risk, WWF, May 2002). Most of these populations are in Canada with the balance scattered around the perimeter of the Arctic Ocean. Up until recently, the Arctic Ocean was not considered a polar bear region. Neither was Greenland.

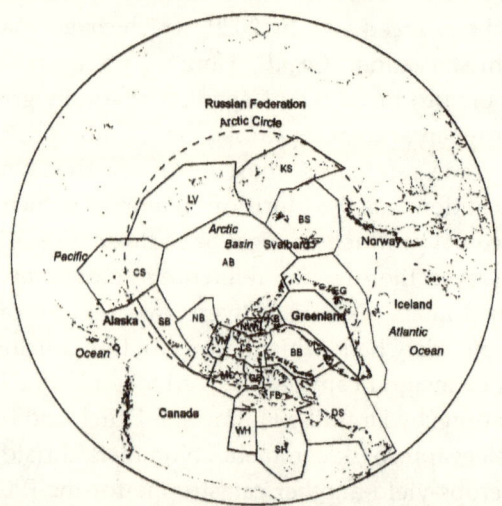

Legend: CS = Chukchi Sea; SB = Southern Beaufort Sea; NB = Northern Beaufort Sea; VM = Viscount Melville Sound, NW = Norwegian Bay; LS = Lancaster Sound; MC = M'Clintock Channel; GB = Gulf of Boothia; FB = Foxe Basin; WH = Western Hudson Bay; SH = Southern Hudson Bay; KB = Kane Basin; BB = Baffin Bay; DS = Davis Strait; EG = East Greenland; BS = Barents Sea; KS = Kara Sea; LV = Laptev Sea; AB = Arctic Basin

Fig 23e: Polar Bear Population and Distribution
(Federal Register, Vol 71, No 5, Page 1069)

In November 2004, the 1,042 page Arctic Climate Impact Assessment (ACIA) Report was issued. It is available free for download from <http://www.acia.uaf.edu/pages/scientific.html>. The conclusion of this report on page 990, section 18.2.1.1 reads: "The climate of the Arctic has undergone rapid and dramatic shifts in the past and there is no reason that it *could* not experience similar changes in the future. Past changes show climatic cycles that *have occurred regularly* on time scales from decades to centuries and longer and are most likely to have been caused by oceanic and atmospheric variability and variations in solar intensity". Read this over and over again. There is no mention of CO2 or temperature increases. The conclusion is: the arctic climate occurs in cycles due to ocean currents, atmospheric currents, and solar intensity.

In the above ACIA report, projections of future climate are made in section 18.2.1.3. Five different computer models were used, building on computer projections made by the IPCC. Oh, just what we needed; more of the same primitive, useless computer models, based on the distorted temperature record, cooked up by Dr. James Totts Hansen.

The entire issue surrounding the polar bears comes down to one thing: OIL, or to be specific, oil that could be harvested from ANWR, the Arctic National Wildlife Reserve. In order to put a stop to drilling at ANWR in Alaska, the environmentalists needed to invent a good reason. Getting the polar bear moved up on the endangered species list from "threatened" to "vulnerable" provided this justification. The polar bear extinction threat has been manufactured not from facts, which show no problem at all, but from faulty computer projections based on a distorted temperature record. And this crap is referred to as "science".

Greenpeace-gate

A recent report was released by the IPCC claiming that by the year 2050, 80% of the world's energy would come from renewables. <www.joannenova.com.au/2011/06/greenpeace-gate-breaks-and-the-ippc-gets-busted...>. And I have the paperwork proving sole ownership of a bridge in Brooklyn I would like to sell you. Steve McIntyre, yes the guy along with Ross McKitrick who exposed the "Hockey Stick" as a fraud, discovered that the lead author of this report was not only a Greenpeace employee, but reviewed his own work. So much for independent thought and peer-review at the IPCC.

Greenland Ice Melting Gate

This just in. New satellite data shows massive new melting of ice is occurring in Greenland in 2012, and the melt season isn't even over. <http://www.nasa.gov/topics/earth/features/greenland-melt.html>. Because of this, sea levels will rise even faster. What more proof do you need that there is global warming?

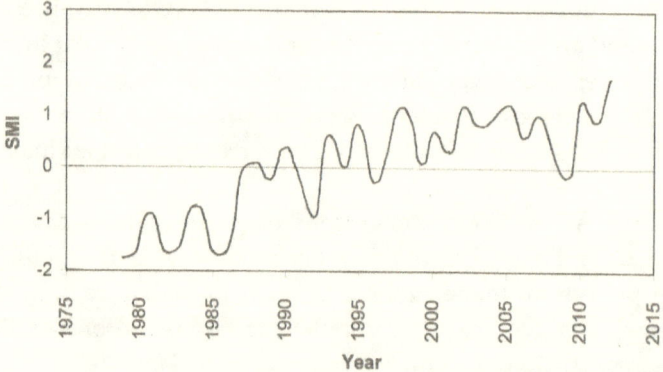

Fig 23f: Standardized Melting Index, 1979-2012

And what is the measurement for this melting? The basis for this claim is the change in the Standardized Melting Index (SMI) between 1979 and 2012. In geological terms this 33 year period would be classified is a pipsqueak of data. But, what exactly is the SMI? It is defined as the number of days when melting occurs times the area of melting. You heard that correctly. The AREA of melting is being used to guess at how much VOLUME of ice is being lost. The thickness of the melting could be a fraction of an inch, even just surface wetness. The volume of melt is unknown, but that won't stop a lot of overpaid, on-the-public-dole scientists from inventing new computer algorithms to convert two-dimensional data into a three-dimensional myth.

The satellite data being gathered only measures area and is incapable of measuring the volume of melt below the surface. Meanwhile, there is no mention of the change in elevation of the ice over the entire continent, which would be very relevant in determining the inventory of ice gained or lost. But to be sure – if there was an elevation loss you would have heard about it by now. So, the only conclusion that can be drawn from this is: There has been no change in inventory at all, or perhaps – fasten your seat belts – there was an increase, as evidenced by the finding of the Glacier Girl.

A group of 15 scientists have analyzed 47 proxy archive records from glaciers, lake, coastal and deep marine sediments, and peat sequences spanning the last 10,000 years. (Briner, J. P., et al, Holocene climate change in Arctic Canada and Greenland, Quaternary Science Reviews, Vol 147, September 2016, Pages 340-364). They conclude that the decrease in Greenland Ice Sheet surface area was 11 times larger 4,000 years ago when compared to the last 100 year change using today as a basis.

So when the warmistas try to scare the daylights out of you about the precipitous decline of ice surface area in Greenland, tell them to get a grip. The change in ice area in the last 100 years was about 9% of what it was 4,000 years ago compared to today.

Holland-Gate

The IPPC AR4 makes the claim that 55% of Holland is under water, when in fact only 26% of the country is under water. This disparity gives a false percentage of area under water by adding the area of 29% threatened by flooding to the 26% area under water. Though this is not a cover-up, it falsely gives the impression that the flooding in Holland is worse than previously thought. ttp://www.google.com/hostednews/afp/article/ALeqM5hHMDg45IUGEUxRnGBRiwWpSxukFg>

Russia-Gate

The Moscow-based Institute of Economic Analysis (IEA) has claimed that the Hadley Center for Climate Research probably tampered with Russian climate data. The IEA said that data submitted by only 25% of the stations were used, and that 40% of the territory was not included. <http://en.rian.ru/papers/20091216/157260660.html>. Why should this be a surprise? The temperature keepers have been tossing out measuring stations around the world by the thousands, and increasingly faking measurement station data.

Pachauri-Gate

Dr. Rajendra Pachauri was the Chairman of the IPCC up until 2015. It also looks like he has been caught with his hand in the cookie jar. Some detractors of Dr. Pachauri have made issue of his previous occupation. In his past, Dr Pachauri was a railroad engineer. I would not question this career change, or his competency to hold his present position. In fact, as you are learning about the workings and efficacy of the IPCC, Dr. Pachauri seems to have all the qualifications to fit right in.

In June 2007, the India Climate Exchange (ICX) was launched. This exchange would be a sister organization of the Chicago Climate Exchange (CCE). The ICX provides a platform for buyers and sellers to trade Certified Emission Reductions (CERs), or carbon trading credits <http://news.oneindia.in/2007/06/15/india-climate-exchange-launched-1181900243.html>. In January 2008, ICX announced it had established a "founder member group" with one of its members as Dr. Pachauri. This puts Dr. Pachauri in a direct conflict of interest, where he is in position to effect policy on carbon trading credits as Chairman of the IPCC and also in a position to profit as head of the market that trades them.

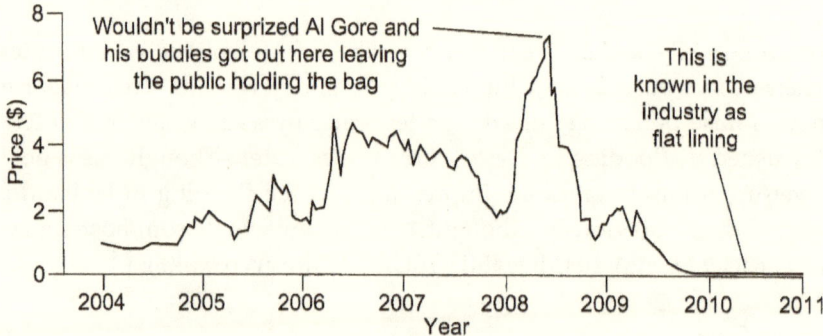

Fig 3g: CCX Contracts Daily Report
(Based on www.chicagoclimatex.com/market/data/summary.jsf)

In January 2004, the Carbon Financial Instrument or CCX contract debuted at $1. For those of you unfamiliar with this trading "instrument", it falls under the general category of a derivative. In June of 2008, CCX peaked above $7.00. By December 2009, CCX trading volume ceased, and around February 2010 the contract price "flatlined". On April 30, 2010, the ICE bought CCX for $600,000,000.

$600,000,000 for what? How can a trace atmospheric gas be monetized? This is just another fractional reserve banking scam like the Federal Reserve System, under whose tutelage the value of the US dollar has decreased 96% since 1913. In the case of ICX, this wasn't the business of greenhouse gas reduction, it was the business of taking suckers to the cleaners.

I don't know about you, but there is an unpleasant aroma wafting around this derivative. This particular carbon trading system was once heralded as a win-win situation for both buyers and sellers. Anyone familiar with the stock market knows the maxim that for every buyer there is a seller, and for every winner there is a loser. By definition, there is no win-win situation in Wall Street. From my perspective a great deal of money was made and lost. The question is: who won, who lost, and how did Dr. Rejendra Pachauri figure into this whole mess?

Conclusion

- The IPCC is the "scientific" arm of the UN.

- The IPCC is considered the 'gold standard' when considering what positions should be taken by policymakers in matters of climate legislation.

- The IPCC should be considered for what it really is: a powerful body of corrupt bureaucrats, often overriding the findings of its own scientists, guided by a Socialist agenda, within a Socialist organization.

Chapter 24: NCAR Assessment

I just can't get enough of these Assessment Reports (ARs). During the last 20 years we have had to endure five ARs from the UN IPCC. As if these UN ARs weren't enough, we have been recently exposed to National Climate ARs. The following is my assessment of the National Climate Assessment Report of 2014.

Executive Summary

The 2014 NCAR was commissioned by President Obama for use by "the Executive Branch to underpin future policies and decisions to better understand and manage risks of climate change." First, it is obvious this report was requested by the President to give him the ammunition for his climate change and sustainability goals. Second, it is pure hubris to assume we mortals can manage climate risk. When the Sun decides to throw a tantrum, even a micro-tantrum, or the Earth experiences a magnetic pole reversal exposing all life to a tsunami of DNA-altering, life-threatening radiation in the process, all this global warming, climate change, climate disruption, climate catastrophe nonsense will flash into obscurity.

The 2014 NCAR is lacking throughout. Within its 16 "our changing climate" sections there are references, many of which appear repeatedly in every section. This bulks up the list of references much like double-spacing a JHS report to make it look more voluminous to the teacher. Further investigation into these references also give evidence contradictory to many of the conclusions reached. This report was prepared "by a team of more than 300 experts guided by a 60-member Federal Advisory Committee". Translation: 300 people with hopefully scientific credentials had their work re-arranged and conclusions revised by a bunch of bureaucrats.

Section By Section Analysis

Introduction Section

The Referenced Period For Graphs of 1901 to 2000 stands out like a sore thumb. Out of millions of years of data, this section refuses to acknowledge anything happening before 1901. More significantly, the section does not acknowledge anything after 2000. This is because the evidence since 1998 shows that warming as measured by satellites has ceased, even while CO_2 concentrations continued to increase. It has become obvious since 1998 that CO_2 is not driving the temperature. In fact, ice core data shows CO_2 increases occur between 600 to 1,000 years AFTER temperature increases. It is warming oceans that release CO_2 out of solution into the atmosphere that is responsible for the CO_2 increases. CO_2 is not a player in climate change, CO_2 is a spectator.

Observed Change Section

In this section two Graphs are presented, one with CO_2 concentration from 1880 to 2014, and a second with temperature during this same period. What is not shown is a temperature graph from 1997 to 2014, which would reveal that there has been no temperature increase.

There exists another CO_2 and temperature resource spanning over five hundred million years, which shows that there is no correlation between CO_2 and temperature at all. These well publicized graphs were conveniently omitted from this section. No reference is made to the this ground breaking and thorough work of R. A. Berner and C. R. Scotese.

Future Climate Change

This section shows the actual temperature from 1900 to 2000, and projected temperature from 2000 to 2100. What is not shown is the failure of all global climate models to predict the lack of warming since 1998. How can it be that every one of the 90 models recognized by the IPCC failed to predict this temperature pause? Easy. If the models are based solely on greenhouse gases with no accounting for the cycles and interaction of the Sun, Moon, and planets the models are doomed for failure at predicting climate.

There is also a problem with the numerous studies commissioned by the likes of the EPA to scientists, if only they can conclude that man is somehow responsible for warming. This can and does result in a self-fulfilling prophecy. If studies are only commissioned to find warming, what else will the conclusions be?

Recent U.S. Temperature Trends.

This section acknowledges temperature back to 1895, a period of time which is infinitesimal and irrelevant in geological terms.

This section does not take into consideration the tampering of temperature data by Dr. James Hansen of NASA-GISS and his cohorts, who have systematically adjusted raw temperature data, which everyone in science and government rely upon for accuracy. Astonishingly, the raw data has been destroyed and cannot be verified. You will just have to "trust" the adjusted data as being accurate. The entire process of temperature data collection and manipulation has been corrupted beyond repair and is nothing more than a pile of garbage created by scientific malpractice.

Frost-Free Season

The span of time used for this section begins 1980. Why would any intelligent human being look at a measly 37 years of recent climate data and attempt to make any reasonable conclusion on this subject? This isn't science. This is a meaningless speculation.

Precipitation Changes

Nowhere in this section is credit given to powerful ocean oscillations which are a prime driver of climate cycles. The Pacific Decadal Oscillation and Atlantic Multidecadal Oscillation are the main drivers of precipitation/drought cycles affecting this country. The Indian Monsoons are predictable and caused by the Indian Dipole Oscillation.

Heavy Downpours Increasing

When the climate gets warmer, the air is capable of holding more water in the form of vapor, which creates more clouds that reflect more energy back to space, resulting in cooling and moisture condensation falling back as rain, completing the cycle. This is the beauty of nature recycling heat in a most efficient manner. We are in a warm-wet phase which started in 1995 and will end in 2019 according to Harris-Mann at Long Range Weather. But, of course, this historical warm-wet, hot-dry, cool-wet, cold-dry 70-120 year climate cycle discovered by Dr. Raymond Wheeler did not make it into this section.

Extreme Weather

This section speaks of heat waves, which are nothing more than blocking high pressure areas held temporarily captive by a wrinkle (Rosby Wave) in the jet stream. Eventually, the jet stream straightens out, the high pressure area moves on, and the heat wave ends. It always does, and always will.

There is evidence that drought cycles in this country have been going on for at least 4,500 years according to Jim Clark at Duke University. But apparently this was not worth mentioning in this section.

Every State has an all time high record temperature. Of the 50 states, 23 of the all-time hottest temperatures were set in the decade of the 1930's. It is acknowledged there were some months of record setting high temperatures in TX, NM, OK and LA. However, the first quarter of 2014 has proven to be the coldest 3 month period in the history of the USA. Unfortunately, this cold temperature record did not get acknowledged in this section.

The recent four decades of warming are in sync with the Grand Solar Maximum that occurred during this period. The Sun has a 360 year cycle of regular oscillations, followed by a grand maximum, followed by a grand minimum. As of 2009, the Grand Solar Minimum has begun. Prepare for decades of bone chilling cold and wet winters.

Changes In Hurricanes

Yes, there have been a couple of outstanding hurricanes and typhoons since 1980. However, the Total Accumulated Cyclonic Energy (ACE) of the world, which is a broad measure of the hurricane and tropical cyclonic energy, is at a 40 year low as monitored by Dr. Ryan N. Maue at Florida State University. With rare exception, the hurricane seasons of the last couple of years in the US have been complete duds.

Hurricane Sandy was really a superstorm, barely a Category 1 hurricane. This storm was predicted a year in advance by Global Weather Oscillations, which predicted a "hybrid hurricane" coming ashore in the Northeast US. Indeed, Sandy was the result of a perfect storm between a Nor' Easter and a hurricane. Good call GWO. No call NCAR.

Then along came the year 2017 and hurricanes Harvey and Irma in quick succession. Yes these were truly devastating storms. Not to sound insensitive, but after 12 years of the lower 48 states not having a major hurricane make landfall, the odds were tilted heavily in something major happening eventually, and it did.

In the case of Harvey the situation was unusual. The storm made landfall, but instead of passing over and moving rapid inland, Harvey decided to stop, turn back, regroup over the edge of the Gulf, reinvigorate itself, and then ever so slowly meander off to the north and east over a period of three days, all the while continuously dumping rain on Houston, a very flat, very large city, covering hundreds of square miles. Over the years Houston was the wild west in terms of unregulated development. Green spaces which normally absorbed rainfall, were covered over with asphalt, concrete, commercial buildings, and housing tracts. Outdated infrastructure, historically incapable of handling even normal downpours from the usual thunderstorms, were no match for Harvey. With so much area and so much rain there was only one place for the water to go – and that was up, causing monumental flooding. Unbelievably, some of the city's major highways were designed to take up the slack in the surge basins paved over by development and provide the emergency storage for heavy rains. That is why news pictures of Houston's highways with cars under water are so common. These highways were designed to be surge ponds for downpours.

In the case of Irma, each year that passed more and more development was placed in harm's way. Florida was living on borrowed time. It was only a matter of when, and Irma answered that question. In retrospect, Florida should consider itself lucky. Can you imagine the wind and tidal surge damage that could have been wreaked had Irma taken a path right up the east coast from Florida, to Georgia and South Carolina? The damage would have been astronomical. Instead Irma headed up the peninsula, weakening quickly, thus sparing Florida from what could have been much worse.

Changes In Storms

Hurricanes and tornadoes in the US are at an all time low. However, there has been increased storm activity in the British Isles and Europe as predicted months in advance by Dr. Piers Corbyn at www.weatheraction.com. As the British Meteorological Office with its super computers consistently misses their predictions, Dr. Corbyn, armed with only a laptop computer, consistently predicts weather months in advance with 85% accuracy.

Sea Level Rise

Nils Axil Morner is the preeminent expert on sea level rise, and has said that there has been no sea level rise at all. Dr. Morner, and his research of course never made it onto the list of references in this section. Dr. Morner has stated that the 2.8-millimeter rise being claimed for the world's oceans hinges on one tidal gauge at Quarry Station in Hong Kong Harbor. The sea is not rising at this station, the station is sinking into the sea.

Melting Ice

This section contains a five decade bar chart showing the decrease in ice extent for the Great Lakes from 1963 to 2013. However, the ice extent of the 2013-2014 winter season for the Great Lakes, which is not part of this bar chart, was a record breaker with over 92% of the Great Lakes frozen over. This fact didn't make it into this section.

Much attention is spent on the Arctic which contains only 0.1% of the world's total inventory of ice. The demise of the Arctic Ice Cap has been an ongoing prediction for the last 50 years, but it continues to survive.

Meanwhile, the Antarctic Ice Sheet, which contains about 900 times more ice than the Arctic, continues to get wider and thicker. Measuring stations are continually getting buried under snow and ice. But no mention is made of the Antarctic in this section.

This section refers to Greenland's melting ice "area". Area of melting with no thickness measurements is meaningless. It is the total ice volume which is important. So, the surface edges of Greenland are melting. Big deal!

Choosing Alaska's glaciers for discussion is cherry picking. For every glacier that is shrinking, there is another glacier that is growing. Some glaciers are large, some are insignificant. And here's a news flash for all warmists out there: we are at the very end of the present 10,500 year Interglacial Warmup Period. During these warmup periods glaciers melt. That is what glaciers are supposed to do during Interglacials. What should be of immense concern is the numerous glaciers that are growing.

What matters is the world's entire inventory of ice. In the absence of any hysteria on this subject, one can only conclude from this and other reports that the world's total inventory of ice is stable or growing. If it wasn't, we would have heard about it by now.

Ocean Acidification

Theoretical pH projections are not borne out by real world habitat measurements. pH changes between 0.024 and 1.42 have been observed out in the field, are highly site dependent, and well tolerated by ocean life. The very small pH changes projected in this report and their effects on ecosystems would be underwhelming compared to actual changes already being observed.

National Climate Assessment Report of 2018

The 2018 NCAR borrows from the same recent myopic time frame of limited data yet manages to expand the previous report of 840 pages into over 1,000 pages. Proponents of AGW complained that the report was released on a Friday before a holiday, which would result in a minimal response from the public.

Conclusion

The 2014 National Climate Assessment Report is 840 pages of the same old global warming bilge, replete with distortions, errors and omissions.

"Because the myth is that promoting science isn't just about providing resources – It's about protecting free and open inquiry. It's about ensuring that facts and evidence are never twisted or obscured by politics or ideology"

– Barack Obama

Departure

<u>Global Warming / Climate Change Myths</u>

Just about all video and pictorial accounts of global warming and climate change events do not take into account that a hundred years ago everyone did not have a smartphone to take HD videos of weather related events. Eighty years ago it was unusual for a tornado to be recorded by someone rich enough to afford a 16-mm camera and its expensive film. Today, a single tornado can be recorded by dozens of individuals possessing a Smartphone from various angles along its entire path of destruction. These events are impressive and appear to be numerous, but they are not unusual and have been going on for millennia. It is only because of today's technology made affordable to almost everyone that these recordings are now so plentiful.

The myths about increasing temperatures, increasing CO2 concentrations, melting ice, decreasing sea ice extent, rising sea levels, stronger storms, more severe droughts, bigger floods, larger forest fires,

Polar Bear extinction, etc., have simply not materialized. These myths have been discussed and disproven by the evidence presented in the previous chapters. Permit me to summarize these myths as follows.

Increasing Temperatures. For the last 600,000,000 years temperatures have hovered around 12C about 14% of the time, around 22C about 50% of the time, and somewhere in between 36% of the time. Right now we are at 14.5C, about 25% above the bottom of the historical range. We are no where near any temperature tipping point.

The 0.4C rise in temperature since the Industrial Revolution (IR) pales in comparison to the 1.6C increase of the Medieval Warming Period (WP), the 2.5C increase of the Roman WP, and the 3.2C increase of the Minoan WP, using the IR as a baseline. The average temperature has been in decline for the last 6,000 years.

There have been three global cooling and three global warming periods within the last 250 years, all marching to the tune of changing solar irradiance, not CO2 concentrations.

Interglacial Warmup Periods (IWP) last about 10,500 years. We are at the very end of the present IWP. After this comes about 90,000 years of snow, ice, advancing glaciers and incredible loss of life. Enjoy the warmth while you can.

Increasing CO2 Concentrations. About 550,000,000 years ago CO2 was 7,000 ppm and has decreased to where it is today, near it's historic low. If CO2 fell below about 100 ppm, photosynthesis would cease and our lush green planet would become a lifeless ball of snow and ice. On the other hand, plants thrive in nurseries kept at CO2 concentrations of 1,000 ppm. Thanks to recent CO2 increases, vegetation has increased 11% in the arid areas of the world.

The famous Mauna Loa CO2 measurements began in 1958, coincidentally at a historic low CO2 level of 315 ppm. In 1942 and again in 1822 CO2 was 440 ppm, 40 ppm higher than today. We are 300 ppm away from a world without vegetation and 6,600 ppm away from a world that existed millions of years ago just fine. To say we are walking in a fog heading for the edge of a cliff, being 4.3% off the bottom of the historic CO2 range is ludicrous.

Antarctic ice core data reveals that the CO2 increases occur about 600-1,000 years AFTER the temperature increases. CO2 is not driving the warming. The warming is driving CO2 out of solution from the world's oceans into the atmosphere. Let me say that again – temperature drives the CO2 changes, not the other way around.

For the last 150 years there has not always been a correlation between fossil fuel use and temperature. Between 1940 and 1970 while CO2 increased, fossil fuel use decreased. If all the world's known fossil fuel reserves were burned overnight, the resulting CO2 temperature increase would be no more than 5C.

Melting ice. Over the last few years there have been unrelenting cries about the melting Artic Ice Cap, melting glaciers, and ice sheets and ice shelves slipping off into the sea all over the world. When you see glaciers calving (ie: ice dropping off at the head of the glacier, usually into a large body of water), it isn't due to global warming or climate change. It is because of the pressure of millions of tons of accumulating snow and ice upstream that is pushing the glacier downstream. What is driving the newsreel footage is not the implications of fossil fuel burning, but the reality of the physics of a plastic mass being pushed toward the sea by increasing snow and ice at higher elevations, far away.

Decreasing sea ice extent. The myth of shrinking sea ice extent is not supported by the evidence. Sea ice extent and volume continues to accumulate at record levels in the Antarctic where 90% of the world's ice inventory is located. Melting of Antarctic ice shelves have been traced to underwater geothermal activity not CO2. The Arctic Ice cap varies from about 5 to 15 million square kilometers in area from summer to winter. It is controlled by a pulse of warm Pacific water entering every 72 years. The Arctic Ice cap is about 3 meters thick and about 10,000,000 square km in area, which is comparable to a piece of paper the size of four football fields. The fact that this fragile, virtually two-dimensional, film of ice doesn't disappear overnight is astonishing.

Rising sea levels. I am really getting fed up with these stupid graphic images of Florida and various US coastlines disappearing when all the world's ice melts. The myth of sea level rise threatening coastal areas has been going on for more than a century. News accounts about islands disappearing underwater, can be traced to drainage of water inland for farming or subsidence of geology. Sea level rise according to Nils-Axil

Morner, the world's leading authority on sea level change, has not changed at all. Besides, we are at the very end of the present Interglacial Warmup Period where it should be expected for ice to melt. What you should be concerned about now is why in other areas of the cryoworld glaciers and ice are growing.

Stronger storms. Annual Accumulated Cyclonic Energy (ACE) is the barometer of how active cyclonic events have been for any year. ACE is at historical lows, as are the overall number and strength of hurricanes and tornadoes. As of 2014, tornadoes and hurricanes are at 40 year lows. The hurricane seasons of 2012, 2013, 2014, 2015 and 2016 in the USA have turned out to be complete duds. In the last half of May 2019, the US Midwest suffered thru a relentless train of tornadoes. This relentless assault was the result of a blocking high pressure area over Florida dumping endless quantities of warm moist air, at the same time a low pressure area over the NW poured endless quantities of cold air, into the same location. Any surprise this confluence should occur in what is rightfully called 'Tornado Alley'?

More severe droughts. Droughts are nothing new. Droughts have been the scourge of past civilizations and resulted in their obliteration. In perspective, the recent drought in California that lasted for several years was a blip in drought history compared to droughts in the past that lasted for generations. Civilizations such as the Anasazi in New Mexico, Minoan and Mycenaean in Greece, Nabta Playa in Egypt, Mesopotamian, Mayan, Khmer and Angkor Wat in Cambodia, and numerous other cultures met their demise from droughts. Droughts in the Great Plains of the United States are also nothing new and continue their march in tune with the 18.6 year lunar cycle. The most severe droughts occurred hundreds and thousands of years ago, well before any influences of man from fossil fuel burning. News of droughts also do not take into consideration the strain put on agricultural land trying to grow food for an exploding world population, massive land clearing campaigns which reduce available land for growing, and poor management of soils that are already depleted.

Bigger floods. The myth of more floods is not supported by the historical data. It also does not take into consideration that over the decades and centuries humanity has encroached more and more towards the waters' edge, thus increasing their risk of being impacted by floods. The top six all time world's worst floods as measured by loss of life occurred in China. Also, five of the ten worst floods occurred before the Industrial Revolution <www.epicdisasters.com>. Recently, it has been discovered that an El Nino type mechanism operates in the Indian Ocean which supplies the monsoon rainfalls of India and East Africa. El Ninos are creations of nature. Mankind has nothing to do with El Ninos coming and going. The news coverage of flooding around the world does not take into consideration the direct influence of man on flood control. These include building levies and dykes which control local flooding but only push the problem further downstream for someone else to handle, land development near rivers or harbors which fill in needed surge volume for rising waters, or land clearing for development which depletes the land's ability to absorb water.

Larger forest fires. Increasing forest fire activity remains at the mercy of lightning strikes, underbrush stockpiles and human interferences. Over the last five decades forest management practices involved putting out smaller forest fires to protect the ever increasing encroachment of human migration into rural areas. This fire fighting philosophy resulted in a build up of small diameter trees and woody debris of unprecedented proportions. The result has been the creation of bigger uncontrolled forest fires, or wildfires <www.americanforests.org>. Only recently, has this philosophy changed back to what nature intended. Let the forest burn. This is nature's way. This replenishes the soil, creating new trees, plants and bushes, which becomes food for wildlife species. Many forest ecosystems are fire dependent, needing fire to germinate and replenish <www.enviroliteracy.org>.

Polar Bear extinction. Polar bear populations – regardless of how many photos are posted of bears precariously perched on tiny ice floes barely large enough to support them - continue to go about their business unphased by supposed climate changes. That is because polar bears live on land, hibernate on land, reproduce on land, and hunt from land near water resources. They do not loiter continuously on pieces of ice the size of a compact car floating out in the middle of nowhere. For the last few decades polar bear populations have remained relatively stable, a news story that gets no respect.

BIG CLIMATE

The myth of global warming, climate change, climate disruption, climate catastrophe, or whatever they are calling it today, continues, because of the trillions of dollars that would be lost and millions of leaf-raking jobs eliminated, if this charade were to be exposed.

-- Banks and brokerage houses reap huge commissions from it. These commissions come from sponsoring various alternative energy projects such as solar power, wind power, biomass, and corn ethanol. If it wasn't for global warming these useless, unsustainable – yes, you heard me right – these unsustainable projects would never have gotten off the ground.

-- Scam artists like Maurice Strong (who passed away in 2015) thrive on it, creating schemes like carbon trading which suck billions of dollars from consumer's wallets. Carbon trading establishes an arbitrary, artificial value to a trace gas which otherwise has no value all, and thru political coercion forces utility companies and industries to buy these "credits" lest they have to shutdown their operations. Like a Ponzi Scheme the first to enter and shortly leave these carbon trading schemes are the first to profit. Those who hang around too long will ride the value of their credits to the eventual, predictable flat line. This scam is nothing more than a tax, which ultimately is passed onto customers as higher electric bills or higher prices for products.

-- Politicians thrive on crises. If there are no crises, politicians will invent them. Politicians need crises to inflate their egos; that they are curing people's problems. They need crises to justify tax increases so they can create and fund entitlement programs that garner votes and keep themselves in office in perpetuity.

-- Bureaucracies need it to grow bigger and more powerful. Sometimes bureaucracies provide funding to further research into global warming. This feeds the fires of hysteria and thus escalate the need for more funding to keep their empires expanding and the bureaucracies growing even further.

-- Many scientists keep busy by grazing at the trough of free grant money made available, only if it can be shown that man causes climate change.

-- Corporations need it to sell cures for which there is no disease, and fatten up their bottom lines. DuPont was all too willing to suspend manufacture of its Freon 12 and 22 refrigerants, and replace them with a whole new product line of equally questionable, supposedly environmentally friendly refrigerants. General Electric cannot wait to make as many wind turbines as possible. Corporations waiting in line for funding of solar panels stretch as far as the eye can see, and the monies garnered quickly disappear like the Solyndra fiasco into a black hole.

-- The alternative energy, Green Building and sustainability industries came into existence and thrive off of it. Every corporation espouses to be "Green", regardless of what product or service they offer.

-- The news media needs it to keep the frenzy going, the ratings up, and ad revenue coming in. The legacy news media thrive on chaos. They are in the sole business of bringing us chaos into our newspapers, magazines and television. If it wasn't for chaos, they would have gone out of business a long time ago.

-- The United Nations needs it to forge its role as the leader in One World Governance. Let us be honest with each other. The UN is a collection of Republics, Democracies, Kingdoms, Monarchies, Theocracies, Dictatorships, and Kleptocracies (a form of political corruption where the government exists to increase its personal wealth and power at the expense of the wider population). The vast majority of UN countries consider themselves to be less fortunate than the United States and are a bunch of whiners.

-- Environmentalists, anti-industrialists, Earth-Firsters, and other Communists need it in order to cut the legs out from underneath the evil, Capitalist United States and level the playing field for the world's less fortunate nations.

"Science is the great antidote to the poison of enthusiasm and superstition"

– Adam Smith

Aftword

We have heard it so many times before, we are getting brain numbed: '97% of scientists agree.' '98% of scientists agree.' 'Ignore the small, radical group of skeptics who are funded by Big oil.' 'The science is settled.' 'The debate is over.' These are just some examples of the sound bites coming from the Main Stream Media, Hollywood celebrities, and global warming acolytes such as Al Gore. The drumbeat of this dogma of global warming is loud, repetitive and never-ending. It is the big lie; say it over and over and over, and it becomes accepted as the truth.

Global Warming Pause

But almost without notice, the argument about global warming has quietly morphed into climate change. This stealthy movement was necessary because the warming as measured by satellites statistically stopped around November 1996, even while CO_2 concentrations have continued to increase. Since then, it has become increasingly obvious CO_2 is not driving the weather, the warming, the climate, or anything else.

During the 1970's the real fear was that fossil fuel burning was responsible for catapulting us into the next 90,000 year Ice Age. A couple of decades later, the Ice Age fear was replaced with the Earth-Is-Turning-Into-A-Cinder from global warming fear. Then in 1996, satellite measurements began indicating no increases in temperature. This has been called by some as a 'pause' in what they hope will turn out to be the resumption of relentless increases in temperatures due to fossil fuel burning. Well, the fossil fuel burning hasn't stopped, so why have the temperatures leveled off? Is it not obvious there is a lack of correlation between fossil fuel burning and temperature increases?

Ice Ages and Interglacial Warmup Periods

Our present civilization of the last 10,000 or so years has prospered only because of the onset of the recent Interglacial Warmup Period, which abruptly terminated the last 90,000 year Ice Age. The attendant warming of the last ten centuries is singularly responsible for the rise of dozens of empires, a growth of billions of people, most of whom are enjoying unprecedented prosperity all due to a radical though not unusual warming of the climate – a warming totally natural, totally expected, and unphased by any human influences. A result of the present IWP is that sea levels have risen about 400 feet to where they are today, an imposing fact that is conveniently swept under the rug by global warming stories involving sea level changes of millimeters per year. The fear of sea levels changing mm/yr has become overshadowed by changes in kilometers of sea level change, a story factor of a million to one. The fact that the beach of New Jersey was located about 150 miles east of the present beach near the edge of the continental shelf escapes the attention of the warming advocates. But an increase in sea level of only a few mm in a year is cause for panic.

Planetary Mechanics – *THE* Driver of Climate Change

Planetary mechanics is the study of orbiting celestial bodies, including changes to the solar system barycenter, spin orbit coupling, and conservation of angular momentum. It is the very interaction of the motion of the planets, Sun and moon which dictate our climate and our weather. This isn't theory. This is astrophysics.

Jupiter, Venus and Earth are rightfully called the Tidal Planets, because they control the Sun's tide and its 11 year sunspot cycle. There are many harmonics of this basic 11 year Schwab cycle. There is the 22 year Hale magnetic cycle. There is the 44 year Solar Conveyor Belt cycle. Every 88 years there is the Gleissberg cycle - an amplitude modulation of Schwab cycles. There is the 208 year deVries cycle. The 1,440 year Bond or Ice Rafting Debris Cycle. The 2,400 year Hallstadt Cycle.

There are numerous other cycles which result from combinations of solar, lunar and planetary cycles. Every 18.6 years there is the Lunar

Tidal Cycle which corresponds to abundance cycles on Earth. About every 60 years there is the Pacific Decadal Oscillation and Atlantic Multidecadal Oscillation, two of the most powerful climate forces on the planet.

Then there is Uranus and Neptune (U-N) with their 178 year orbit beat cycle. The Sun also operates in 360 year cycles, a harmonic of the U-N cycle. Each 360 year cycle is composed of Regular Oscillations, followed by a Grand Solar Maximum, followed by a Grand Solar Minimum. This totally predictable 360 year cycle has resulted in the Oort, Sporer, Maunder, Dalton and numerous other unnamed Minimums within the past two millennia.

In 2009, we entered the next Grand Solar Minimum - the Landscheidt Minimum. This isn't unfounded speculation. This is traceable, predictable planetary mechanics. From this point forward be prepared for relentlessly colder winter temperatures which will reach its worst around 2040. Along the way there will be ever-increasing fuel scarcity, crop failures, food shortages, famines and the loss of life into the millions. The next Little Ice Age is just around the corner. No amount of pithy CO_2 increase is going to provide enough life-saving warmth. Prepare for decades of bone-chilling colder winters.

Planetary mechanics is the elephant in the room of climate change. The planets control the climate of the Sun which, combined with the Moon, control the climate of the Earth. CO_2 is only a flea on the elephant's ass coming along for the ride.

The Finale

Let's get to the point of this whole global warming, climate change matter. Yes, manmade CO_2 contributions do have a measurable, although insignificant, effect on climate. Carbon dioxide emissions are not responsible for global warming, global cooling, climate change, weather, or anything else. Blaming man for changes in the climate is like blaming the flies for the smell in the elephant house at the zoo. As you have discovered in the chapters of this book, CO_2 is not a driver of climate change; CO_2 is only a passenger coming along for the ride.

All this climate change baloney is nothing more than a scheme to make you feel guilty for having allegedly bashed the planet, wantonly plundered its resources, and having caused irreversible environmental damage. This guilt injection is necessary to more easily separate you from your money. Its purpose is to redistribute the wealth of richer countries to poorer countries. Its goal is to replace Capitalism with Socialism.

For the most part, this scam has been hiding from public view. Up until now, behind the Green Curtain, forces have been at work disguising this goal of Socialism by creating the concept of a Sustainable Planet. This sustainability crap has its roots in the teachings of Marx and Engels. However, since the big lie of GW/CC is so pervasive, they have no problem now of revealing their true intentions.

Recently, this scam has come out from behind the Green Curtain. The people behind this scam do not have to hide, anymore. For example, Christiana Figueres, the Executive Secretary of the UN Framework Convention on Climate Change stated, at a news conference in Brussels the week before February 10, 2015, "that the goal of environmentalists is not to save the world from ecological calamity, but to destroy Capitalism."

Also, on January 24, 2014, the House Oversight and Government Reform Committee released the deposition transcript of former Senior EPA official John Beale. In this deposition, Beale revealed the ultimate goal of the EPA was to "modify the DNA of the capitalist system." Between 2008 and 2016 the EPA declared open war on Capitalism, in general, and on the coal industry and coal-fired power plants, in particular.

So, like the Wizard in Oz saying to Dorothy, the Tin Man, the Strawman and the Lion, 'pay no attention to that man behind the green curtain'. If you wish, go on with your life and ignore the preceding chapters of hard data, facts, graphs, historical perspective, and logic. You can just keep following the crowd as it is led around by their noses, listening to the global warming and climate change baloney of Big Climate, headline-seeking politicians, and the 'if-it-bleeds-it-leads' mentality of the MSM.

Or – if you dare - you can think for yourself.

Glossary

Albedo – a non-dimensional, unitless quantity that indicates how well a surface reflects solar energy.

Anthropogenic – of, relating to, or resulting from the influence of human beings on nature (then why don't they just say manmade and stop pretending to know more than we do?)

Barycenter – the center of mass of two or more bodies, usually bodies orbiting around each other, such as the Earth and the Moon.

Coriolis Effect - an effect whereby a mass moving in a rotating system experiences a force (the *Coriolis force*) acting perpendicular to the direction of motion and to the axis of rotation. On the earth, the effect tends to deflect moving objects to the right in the northern hemisphere and to the left in the southern and is important in the formation of cyclonic weather systems.

Denier – (Dee – nigh' – er) - a person who knows global warming and climate change are natural, and who refuses to believe in the dogma of the Church of Global Warming and Climate Change.

Dogma - an arrogant and vehement expression of opinion; asserting a matter of opinion as if it were fact; directly affirming rather than being qualified, debated, or discovered by induction; for proceeding from *a priori* truths or assumptions rather than empirical evidence.

Engineering - the application of science and mathematics by which the properties of matter and the sources of energy in nature are made useful to people.

Magnetohydrodynamics – branch of physics which studies the behavior of electrically conducting fluid acted on by a magnetic field.

Peer Review – 1. a process by which something proposed (as for research or publication) is evaluated by a group of experts in the appropriate field. 2. when one member of the 'good-ole-boys club' is asked to review the scientific paper of another member of the 'good-ole-boys' club.

Poli - gray or matter of the brain   **tics** - convulsive motions
Politics – convulsions of the brain.

Policy - art or science of government **maker** –creator
Policymaker – a creator of government.

Proxy Data - preserved physical characteristics of the past that stand in for direct measurements (which were not available in the past).

Science - observation and classification of facts and establishment of verifiable laws by induction and hypothesis.

Warmista - a global warming fanatic

BIBLIOGRAPHY

Cases

agupubs.onlinelibrary.wiley.com/doi/full/10.1002/palo. 20042 113

Alley R. B., The Younger Dryas cold interval as viewed from central Greenland, Journal of Quaternary Science Reviews, 19:213-226 69

Angell, J.K., On the Relation between Atmospheric Ozone and Sunspot Number, Journal of Climate, 01 Nov 1989 21

Archibald, Herbert L., Wildlife Society Bulletin, Vol. 5, No. 3, 1977 153

Balling and Cervany, Influence of Lunar Phase on Daily Global Temperature, Science, Vol. 267, No. 5203, 1481-1483, 1995 155

Beck, Ernst-Georg, Energy & Environment, Vol 18, No. 2, 2007 41

Benuzzi, Eugene Joseph, Masters Thesis: Chandler period atmospheric oscillations at the 700 Hecto Pascal level over the Northern Hemisphere, 20 Dec 1978 94

Berner and Kothavala, AJS, **301**, February 2001, page 197 42

Bollenger, Clyde J., A 44.77 Year Jupiter-Venus-Earth Configuration Sun-tide Period in Solar Climatic Cycles, Academy of Science For 1952. 172

Brauer et al, Nature Letters, **Vol 1**, 520-523, August 2008 95

Braun et al, Nature Letters, **438**, 208-211, 10 November 2005 173

Bray J. R., 1974, Volcanism and glaciation during the past 40 Millennia, Nature **252**: 679 199

Briner, J. P., et al, Holocene climate change in Arctic Canada and Greenland, Quaternary Science Reviews, Vol 147, September 2016, Pages 340-364 237

Broecker, W. S., The great ocean conveyor, *Oceanography* 4(2):79–89, https://doi.org/10.5670/oceanog.1991.07 85

Bryant, Edward, Natural Hazards, Cambridge University Press, Second Edition, 2005 30

Chambers, D. et al, Is there a 60-year oscillation in global mean sea level?, GRL, Vol 39, No 18, 2012 111

Cook, J et al. 'Quantifying the consensus on anthropogenic global warming in the scientific literature.', Environmental Research Letters 2013; 8 024024 220

Crawford, E. A., The Lunar Garden: Planting by the Moon Phases, Capital Books, 2000 155

D. N. Moss, The Limiting Carbon Dioxide Concentration for Photosynthesis, Nature **193**, 587, 10 Feb 1962 73

Deke Xu et al, Synchronous 500-year oscillations of monsoon climate and human activity in Northeast Asia, Nature Communications volume 10, Article number: 4105 (2019) 113

Richardson, John d., et al, Center for Space Research, MIT Cambridge 138

Rieger et al, Nature, 312, 623, 1984 138

Ring, Ken, Predicting the weather by the moon, Gothic Image
Publications, Glastonberry, England, 2000 154

Royal Astronomical Society, 1978, **184**, 759-767 138

Russell et al, A 725 yr cycle in the climate of central Africa during the late
Holocene, Geology, August 2003, v.31, no.8, p 677-680 113

Scafetta, Nicola, Climate Change and Its Causes, ArXiv:1003. 1554v1
[physics.go-ph] 8 Mar 2010 106

Shackleton, Nicholas J., The 100,000-Year Ice Age Cycle Identified and
Found to Lag Temperature, Carbon Dioxide, and orbital Eccentricity,
Science, Vol. 29, No. 5486, 2000 159

Smith, C.; Messina, D., In: Sun and planetary system; Proceedings of the
Sixth European Regional Meeting in Astronomy, Dubrovnik,
Yugoslavia, October 19-23, 1981. (A82-47740 24-89) Dordrecht, D.
Reidel Publishing Co., 1982, p. 39-43. NASA-supported research. 95

Stothers, R.B., 1989: Volcanic eruptions and solar activity. J. Geophys.
Res., **94**, 17371-17381, doi:10.1029/JB094iB12p17371 81

Svensmark H and Calder N, The Chilling Stars, Icon Books, Canada, 2007
.... 80

Tarasova and Fomin, Journal of Applied Meteoroogy, Vol 39, Issue 11,
Nov 2000 20

Ulrich, Roger K., The Astrophysical Journal, 162: 993-1002, December
1970 137

USGS Circular 1254 34

Usoskin et al, 2007, "Grand minima and maxima of solar activity: new
observational constraints", Astron. Astrophys, 471: 301–9 148

Velesco VM et al, ICRC,**1**,553-556 146

Vermeersen and Sabadini, Polar Wander, Sea Level Variations and Ice
Ages, Surveys in Geophysics, Vol. 20, No. 5, 415-440, 1999 159

wattsupwiththat.com/2016/11/24/the-bray-hallstatt-cycle/ 149

Williams, G. E., Solar Affinity of Sedimentary Cycles in the Late
Precambrian Elatina Formation, Aust. J. Phys., 1985, 1027-43 148

Wu, J., Yu, Z., Zeng, H. et al. Possible solar forcing of 400-year wet–dry
climate cycles in northwestern China. Climatic Change 96, 473–482
(2009). https://doi.org/10.1007/s10584-009-9604-4) 112

www.acia.uaf.edu/pages/scientic.html 235

www.brnurosci.org/co2.html 72

www.dieholdfoundation.com/collecting-samples-of-sediments. html .. 38

www.epicdisasters.com 34

www.eurochlor.org/rcei 22

Figures

Acronyms & Abbreviations

ACE	Accumulated Cyclonic Energy	34
ACF	A Convenient Fabrication	204
AGW	Anthropogenic Global Warming	72
AIT	An Inconvenient Truth	9
AIRS	Atmospheric Infrared Satellite	48
AMS	American Meteorological Society	58
AWS	Automatic Weather Station	60
C	Degrees Centigrade	--
CC	Climate Change	93
CERN	European Organization for Nuclear Research	82
CFC	Chloro Fluoro Carbon	23
COM	Center of Mass	148
CRF	Cosmic Ray Flux	80
CRN	Climate Reference Network	58
CRU	Climate Research Unit of the University of East Anglia	50
CV	Circumpolar Vortex	14
DMI	Danish Meteorological Institute	68
ENSO	El Nino Southern Oscillation	35
ESRL	Earth Sciences Research Laboratory	44
F	Degrees Fahenheit	--
GHCN	Global historical Climatology Network	50
GCM	Global Climate Model	69
GHG	GreenHouse Gas	78
GISS	Goddard Institute for Space Studies	47
GMT	Global Mean Temperature	196
GOES	Geostationary Operational Environmental Satellite	33
GRL	Geophysical Research Letters	97
Gt	Gigatons	30
GW	Global Warming	--
Had	Hadley Center of the UK Meteorological Office	50
IPCC	International Panel on Climate Change	45
IWP	Interglacial Warmup Period	72
MSM	Main Stream Media	34
MPH	Mobile Polar High	15
NAO	North Atlantic Oscillation	35
NASA	National Aeronautics and Space Administration	24

Prefixes & Symbols

Kilo   thousands 1E3
Mega   millions 1E6
Giga   billions 1E9
Terra   quadrillions 1E12
Peta   pentillions 1E15

6E9   This number is 6,000,000,000. Notated in engineering units it is represented by the number 6 raised to the Exponent 10 to the 9th power. $6 \times 10^9 = 6E9$.

\#   pounds
\>   greater than
\<   less than
^   raised to the power of
*   multiplication
/   division

Further Reading

Books

Opalek, Charles S., <u>A Convenient Fabrication, The Non-Crisis Of Manmade Global Warming and Why We Are Powerless To Change The Climate</u>, (www.lulu.com, 2007)

Opalek, Charles S., <u>Wind Power Fraud, Why Wind Won't Work</u>, (www.lulu.com, 2010)

Dewey, Edward R., <u>Cycles – The Mysterious Forces That Trigger Events</u>, Hawthorne Books, New York, 1971.

Landscheidt, Theodor, <u>Sun-Earth-Man, a Mesh of Cosmic Oscillations</u>, Urania Trust, 1989.

Bryant, Edward, <u>Natural Hazards</u>, Cambridge University Press, 2005.

Rampino et al, <u>Climate – History, Periodicity and Predictability</u>, Van Nostrand Reinhold, New York NY, 1987.

Easterbrook, Don J., <u>Evidence-Based Climate Science</u>, Elsevier, 2016

Felix, Robert W., <u>Not by Fire but By Ice</u>, Sugarhouse Publishing, Bellevue WA, 2005.

Websites

<u>www.climateaudit.org</u>, Stephen McIntyre. The man who, together with Ross McKitrick, exposed the Hockey Stick Fraud.

<u>www.wattsupwiththat.com</u>, Anthony Watts, The world's most viewed site on global warming and climate change

<u>www.icecap.us</u>, Joseph D'Aleo. The International Climate and Environmental Change Assessment Project.

Index

Consider the second book by Charles S. Opalek, PE

WIND POWER FRAUD – WHY WIND WON'T WORK

This book exposes the utter uselessness of wind power, including how:

- Wind turbines rarely produce their advertised full power. On average wind turbines only produce about 20% of their nameplate rating.
- Wind power is unreliable and undispatchable. When wind power is needed most, it will likely be unavailable.
- Wind turbines are not environmentally friendly. They are noisy, unsightly, kill bats and birds, interfere with radars, and have been shown to be responsible for a slew of health problems.
- Wind turbines consume electricity whether operating or not. Often this power is not even metered. Care to guess who is paying for this electricity?
- In theory, if 20% of US electric generation was replaced by wind power, the decrease in CO_2 emissions would be an unnoticeable 0.00948%.
- In reality, wind power doesn't reduce CO_2 emissions, because backup power plants have to cycle wildly trying to keep up with erratic wind power output.
- Wind power is so unreliable Germany estimates that by 2020 up to 96% of its wind power capacity will need to be backed up by new coal fired power plants.
- Wind power will not reduce the US's dependency on foreign oil. If wind power replaced 20% of US electric generation, the resulting decrease in oil imports would be a measly 0.292%.
- Wind turbines have an embarrassingly low and unsustainable Energy Returned On Energy Invested value of 0.29. The installation of wind power facilities will consume about 3 times more energy than they will ever produce.

Developers, land owners, banks, brokerage houses, governments, manufacturers, law firms, the green movement, environmentalists, researchers, academia, and the legacy news media will be the big wind power winners. Taxpayers and electric bill payers will be the big losers.

Available at www.lulu.com, Amazon, Barnes & Noble.

Consider the first book by Charles S. Opalek, PE

A CONVENIENT FABRICATION -
THE NON-CRISIS OF MANMADE GLOBAL WARMING AND WHY WE ARE POWERLESS TO CHANGE THE CLIMATE
is the definitive response to the errors and omissions of

AN INCONVENIENT TRUTH - THE PLANETARY EMERGENCY OF GLOBAL WARMING AND WHAT WE CAN DO ABOUT IT

This book addresses point by point the errors and omissions of <u>An Inconvenient Truth</u>. This book knocks out all the legs from underneath the manmade global warming nonsense including how:

- the Mauna Loa station is measuring natural CO_2 emitted by the surrounding Pacific Ocean and the volcano it sits on, not manmade emissions.
- on three occasions within the last 200 years, CO_2 levels have been higher than they are today.
- the pre-industrial revolution baseline of CO_2 adopted by the International Panel on Climate Change is bogus.
- CO_2 concentrations have been 20 times higher within the last 600 million years, while the Earth's temperature has remained within a 10 degC band.
- 95% of all atmospheric CO_2 comes from natural sources.
- 99.999% of all water vapor is from natural sources.
- water vapor by a factor of 26 contributes more to global warming than carbon dioxide.
- man's global warming contribution is 0.28%, nature's is 99.72%.

Global warming is Big Business. The news media, corporations, researchers, bureaucrats, tax collectors, governments, rip-off artists, and entrepreneurs are lining up in a profit-feeding frenzy from this contrived non-crisis. And the poster-boy for this charade and chief benefactor is none other than Al Gore.

Available at www.lulu.com, Amazon, Barnes & Noble.

www.ingramcontent.com/pod-product-compliance
Lightning Source LLC
Chambersburg PA
CBHW031827170526
45157CB00001B/214